I0038245

CTS Mathematics

Revision Guide from the Teacher's Desk for Secondary Schools

by

Michael Boyce, Alexandra School
Renaldo Holder, Alexandra School
Tameisha Cozier, Alleyne School
Stuart Mayers, Combermere School
Richard Forde, Ellerslie School
Shanielle Small, Frederick Smith School
Jerome Stuart, Graydon Sealy School
Patrick Cadogan, Harrison College
Corey Worrell, Lester Vaughan School
Phillip Clarke, Parkinson Memorial School
A'ja Maxwell, Queens College
Clare Williams, Queens College
Rosalind Atwell, Springer Memorial School
Kemar Trotman, The Lodge School
Charles VanderPool, The Lodge School
Lawrence Bishop, The St. Michael School

Editor-In-Chief

Dr. Janak Sodha
Head of Department & Senior Lecturer
Department of Computer Science, Mathematics & Physics
University of the West Indies, Barbados.

CTS Mathematics : Revision Guide from the Teacher's Desk for Secondary Schools

First published 2016

ISBN 978-0-9928510-4-0

Questions   170

Chapter 8   175
Relations, Functions and Graphs   175

Chapter 9   212
Statistics   212

Caribbean Teachers Series (CTS)
Preface

The production of indigenous text represents the coming of age of our teachers and lecturers. This publication of a Mathematics Textbook for secondary schools is a timely and important insertion into our education sector. The collaboration between the Ministry of Education, Science, Technology and Innovation, teachers in our secondary schools and the University of the West Indies, Cave Hill Campus, presents an excellent opportunity for educators to collaborate in lifting the standard of Mathematics education in Barbados and the wider Caribbean.

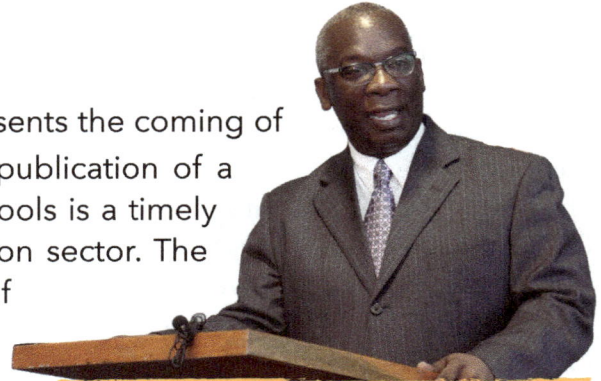

Hon. Ronald D. Jones

Ministry of Education, Science, Technology and Innovation

Barbados

I encourage the students of Barbados and the wider Caribbean to use this important resource to improve their mathematical ability and knowledge. Our societies will not go through the necessary transition to scientific and technologically-led societies without early initiation in Mathematics, Science and Technology. I recommend this book to all students.

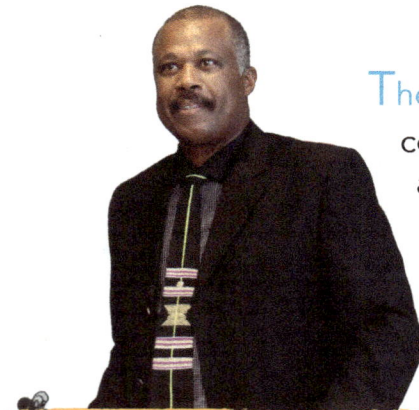

Sir Hilary Beckles

Vice Chancellor

University of the West Indies

The University of the West Indies Cave Hill Campus congratulates Dr. Janak Sodha, Editor-In-Chief and the authors from secondary schools in Barbados on the publication of the Caribbean Teachers Series (CTS) Mathematics: Revision Guide from the Teacher's Desk for Secondary Schools.

Over the last forty years, we have seen the rapid growth of technology and with it the demand for persons who are competent in both mathematics and the sciences. At the same time, we are finding that a large number of young people, worldwide, are struggling with mathematical concepts. While we may all agree that most people can succeed without a degree in mathematics, a lack of mathematical competency significantly diminishes a person's opportunities. We know also, that if the Caribbean region is to take its place in an increasingly technological world, it is imperative that the mathematics and science proficiencies of our young people must be significantly improved and in the shortest time possible.

Written by practising master teachers, who are at the forefront of math's education at the tertiary and secondary school levels in Barbados, CTS Mathematics Revision Guide from the Teachers Desk for Secondary School with its multimedia support is an invaluable tool for students and teachers.

Once again, we congratulate the team of experts on their dedication to the development of their students and especially on the publication of this important addition to mathematic education in the Caribbean

Welcome to your first book in the Caribbean Teacher's Series (CTS). Everyday throughout our schools, teachers create lessons, solve examples on the blackboard to illustrate important concepts but ultimately, rub-off the blackboard! Working alone, a teacher is unaware of how a topic is taught in other schools or which examples are used to highlight key concepts. Furthermore, students are taught a subject by a single teacher, who has no time to write a book.

What is the solution?

Janak Sodha

Founder & Editor-In-Chief
Caribbean Teachers Series (CTS)

University of the West Indies, Barbados

Solution: "Many hands make light work"

By segmenting the exam syllabus into ten chapters and distributing this workload over many different teachers, the Caribbean Teachers Series (CTS) aims to present a window into the knowledge and experience of our teachers for the benefit of our students throughout the Caribbean. A chapter contains **descriptive titles** to enable a student to quickly focus on a given concept. Under each title, theory is presented in a concise manner and illustrated through examples. Every question at the back of the chapter has a video solution, created by the teacher who wrote the question and is viewed through the "**CTS Maths**" app available for your smartphone or tablet.

Is this a book or a virtual school?

> ‣ This book will evolve into essentially a virtual school with teachers from around the Caribbean. Every descriptive title (within the table of contents and highlighted with a green bar in a chapter) will be linked to a video in which a teacher presents the material on a whiteboard to teach the material from fundamentals, assuming no prior knowledge.
>
> ‣ A unique learning experience in which a student essentially attends a **virtual school** through textbooks supported by videos.
>
> ‣ A single subject taught by many teachers, with a focus on the final Caribbean examination for secondary schools.

Chapter 1
Number Theory

Phillip Clarke

Parkinson Memorial School

Barbados

From the very beginning of our academic life, we have come into contact with numbers. We were taught how to recognise different types (sets) of numbers, how to perform simple operations with them, and during this process many of us have asked....why? The answer to theses questions is the basis of number theory, because if Mathematics is the earth, then number theory is definitely its core.

Number theory is important in all areas of Mathematics and other subject areas which require a strong Mathematical base. In fact a good understanding the concepts of number theory is extremely important at the university level where in some cases more letters are seen in some courses than numbers.   This chapter serves to help the student become more comfortable in this topic up to the secondary school level and serve as a launching pad for improvement in other areas of Mathematics and more advanced courses of study.

Recognition of a type of number

Operations performed on a type of number must be carefully examined since they may or may not produce the same type of number.

- **Natural numbers**, N = {1, 2, 3, 4, 5, 6, 7,……..}

- **Whole numbers**, W = {0, 1, 2, 3, 4, 5, 6, 7,……..}

- **Rational numbers**, Q = {…….., $-\frac{1}{2}$, $-\frac{1}{4}$, 0, $\frac{1}{4}$, $\frac{1}{2}$,………}, any number that can be written as a fraction, with a non zero denominator.

- **Irrational numbers**, F = $\left\{\sqrt{2},\sqrt{3},\sqrt{5},\pi\right\}$, any number which cannot be written as a repeating or terminating decimal.

- **Integers**, Z = {…..-4,-3,-2,-1, 0, 1, 2, 3, 4, ……..}, the rational and irrational numbers are not included.

- **Real numbers**, R = {….-4, -3, -2, -1, 0, 1, 2, 3, 4,…..}, the rational and irrational numbers are included here.

- **Prime number**, P = {2, 3, 5, 7, 11, 13,……}, these numbers have only two factors, one and itself.

- **Composite numbers**, {4, 6, 8, 9, 10, 12, 14,15,……}, these numbers have more than two factors.

- **Square numbers**, {0, 1, 4, 9, 16, 25,……}, these numbers occur when an integer is multiplied by itself.

- **Odd numbers**, {……, -3, -1, 1, 3, 5,7,……}, these numbers are not divisible by 2.

- **Even numbers**, {-4,-2, 0, 2, 4,…..}, these numbers are divisible by 2.

Examples

1.   If p , q and r are integers, which calculation below would not necessarily equal an integer?
A. $r(p+q)$
B. $p+q+r$
C. $\frac{q+r}{p}$
D. $p-q-r$

The solution is C because division by an integer, unless the numerator is a multiple of the denominator, may produce a rational number.

2. If a and b are prime numbers then $a \times b$ is
A. Prime
B. Square
C. Composite
D. Even

The solution is C because by multiplying the two prime numbers they themselves become factors of the number produced.

3. Which of the following numbers is irrational?

A. $\dfrac{4}{5}$

B. $\dfrac{\sqrt{2}}{1}$

C. $\dfrac{2}{3}$

D. $\dfrac{1}{4}$

> Each type of number has properties which cause it to be identified.

The solution is B since the root of 2 is neither a terminating nor recurring decimal, and this is the property of an irrational number.

4. Which of the following is a prime number?
A. 441
B. 442
C. 447
D. 449

The solution is D, because if the process of elimination is used, B is even, while summing the digits of A and C both produce a multiple of 3, thus they are divisible by 3.

Find the missing term of a sequence

- Sequences can be in ascending, descending or alternating order.
- They can be comprised of terms which have the same properties (number types) or they can be generated by applying a rule to the first term.

Example

Find the missing numbers in each of the sequences below.

(a)   1, 4, 9, 16, 25 ….

The numbers above are square numbers in ascending order, therefore the missing term would be 36.

(b) $7\frac{2}{5}, 6\frac{4}{5}, ..., 5\frac{3}{5}, 5.$

> While the missing members of a sequence can be sometimes found by applying an operation to the previous term of the sequence, this does not produce the general rule for the sequence.

$7\frac{2}{5} - 6\frac{4}{5} = \frac{3}{5}$ and $5\frac{3}{5} - 5 = \frac{3}{5}$

therefore $6\frac{4}{5} - \frac{3}{5} = 6\frac{1}{5}$, which is the missing term

(c)   4, 6, 9, 13, 18, ...

The sequence is expanding and a simple calculation would reveal that the difference between each term increases by 1, thus the missing number is 24.

Finding the rule of a sequence

The rule of a sequence is generally an algebraic expression in terms of n, where n is the nth term. For example, for the sequence 3,7, 11,15,19,... the rule is not 'add 4' to the previous term, but $4n - 1$.

Example

Find the rule which generates the following sequence, and find the 10*th* term.
1, 4, 7, 10, 13, 16,.......

Term n	1	2	3	4	5	6
Value	1	4	7	10	13	16

Since the difference is consistently 3 in the values 3,6,9,12,15,18 when 2 is subtracted from each value, the sequence 1, 4, 7, 10, 13, 16 is produced. Thus the rule is $3n-2$. The 10th term of the sequence is therefore $3(10)-2=28$.

Use the H.C.F and L.C.M to solve real world problems

For large numbers, it may be safer to use division by prime factors to find both the H.C.F and the L.C.M.

Example

1.   Find the H.C.F of 32, 48 and 96.

2	32		2	48		2	96
2	16		2	24		2	48
2	8		2	12		2	24
2	4		2	6		2	12
2	2		3	3		2	6
	1			1		3	3

Sometimes the H.C.F of a group of numbers is 1.

$32 = 2 \times 2 \times 2 \times 2 \times 2 = 2^5$

$48 = 2 \times 2 \times 2 \times 2 \times 3 = 2^4 \times 3$

$96 = 2 \times 2 \times 2 \times 2 \times 2 \times 3 = 2^5 \times 3$

2^4 is the highest common term since it is the highest power of 2 that is common to all three numbers.

2.   Three ships A, B and C broadcast their location every 15, 16 and 18 minutes respectively. If they last broadcast together at $1:30$ hrs, at what time will they do so together again?

Dividing by prime factors.

3	15		2	16		2	18
5	5		2	8		3	9
	1		2	4		3	3
			2	2			1
				1			

1 hr = 60 min

$15 = 3 \times 5$

$16 = 2 \times 2 \times 2 \times 2 = 2^4$

$18 = 2 \times 3^2$

Taking the highest order of each term gives $3^2 \times 2^4 \times 5 = 720$ min $= 12$ hrs.

$01:30hrs + 12:00hrs = 13:30$ hrs or equivalently, $1:30$ p.m. that the ships will broadcast at the same time again.

Find the value of a digit in any base <= 10 by converting to base 10

- Any number can be converted to base 10 by multiplying each digit by the power of the base it is in and adding them.

- Any number can be converted from base 10 by dividing it by the base you are converting it to and using the remainders.

- The last remainder is the first value.

Examples

1.   The value of 3 in 19374.1 is
A.   3 hundreds
B   3 hundredths
C.   3 thousands
D.   3 ten thousands

> These calculations start from the right (last number).

The number above is in base 10 and **expanding the value would give**
$(1 \times 10^{-1}) + (4 \times 10^0) + (7 \times 10^1) + (3 \times 10^2) + (9 \times 10^3) + (1 \times 10^4)$

Thus the value of the 3 is 3 hundreds, so the answer is A.

2.   Convert 1011011_2 to base 10.

The number is in base 2 and expanding the value would give
$(1 \times 2^0) + (1 \times 2^1) + (0 \times 2^2) + (1 \times 2^3) + (1 \times 2^4) + (0 \times 2^5) + (1 \times 2^6)$
$1 + 2 + 0 + 8 + 16 + 0 + 64 = 91$ or 91_{10}

3.   The 3 in 305_8 is equivalent to what value in base 10?

Expanding in base 8 would give
$(5 \times 8^0) + (0 \times 8^1) + (3 \times 8^2) = 5 + 0 + 192$

Thus, the value of the 3 is equivalent to 192 in base 10 .

> When converting a non base 10 number to a base other than 10, you should convert to base 10 first.

4.   Convert 381 to base 7 .

Divide through by 7.

7	381	
7	54	Rem 3
7	7	Rem 5
7	1	Rem 0
	0	Rem 1

$= 1053_7$

2.   Convert 245_6 to base 7

Converting to base 10 first gives $(5\times6^0)+(4\times6^1)+(2\times6^2)=5+24+72=101_{10}$
Now we convert to base 7 as follows:

7	101	
7	14	Rem 3
7	2	Rem 0
7	0	Rem 2

Thus $245_6 = 203_7$

The commutative, associative and distributive laws

- The Commutative laws for addition and multiplication.
$a+b=b+a$
$a \times b = b \times a$

- The Associative Laws for addition and multiplication.
$a+(b+c)=(a+b)+c$
$a \times (b \times c)=(a \times b) \times c$

- The Distributive law
$$a(b+c) = ab + ac$$

Examples

1.   -pq = -qp is an example of the
A.   Commutative law
B.   Distributive law
C.   Associative law
D.   Inverse law

The answer is A since $-p \times q = q \times -p$.

2. $\dfrac{1}{2}\left(\dfrac{x}{4} + \dfrac{y}{8}\right)$ is equal to

A. $\dfrac{x}{6} + \dfrac{y}{4}$

B. $\dfrac{x}{2} + \dfrac{y}{4}$

C. $\dfrac{x}{8} + \dfrac{y}{16}$

D. $\dfrac{x}{6} + \dfrac{y}{10}$

Using the distributive law gives
$$\frac{1}{2}\left(\frac{x}{4} + \frac{y}{8}\right) = \left(\frac{1}{2} \times \frac{x}{4}\right) + \left(\frac{1}{2} \times \frac{y}{8}\right)$$

$$= \frac{x}{8} + \frac{y}{16}$$

The order of operations

The order of operations is **as follows**:

Brackets   ( )
Division   ÷
Multiplication   ×
Addition   +
Subtraction   −

Remember BODMAS

Example

Calculate $9-(3 \times 2)^2 \div 4 + 2$

Calculating the term in the brackets first	$9-(6)^2 \div 4 + 2$
Squaring	$9 - 36 \div 4 + 2$
Dividing	$9 - 9 + 2$
Adding	$9 - 7$
Subtracting	2

> Note we add -9 and 2, not just 9 and 2. The sign cannot be ignored.

Solve problems involving concepts in number theory

Algebra
- When simplifying an algebraic fraction the L.C.M of the denominators is found in order for the numerator to be simplified.
- When using the elimination method to solve a pair of simultaneous equations, the L.C.M of the coefficients of the variable to be eliminated is found.
- When factorising a linear expression the H.C.F of the terms which comprise the expression must be found.
- When factorising a quadratic expression the factors of the product of the coefficient of the squared term and the constant are used.

Set Theory
- A set is often comprised of a number of terms with specific properties, e.g. odd, even, prime, factors and multiples.

> The concept of a multiple must be understood.

Vectors
A vector is parallel to another if one is the scalar multiple to the other.

Examples

1.   Simplify $\dfrac{3x+1}{2} - \dfrac{5x-2}{4}$

The L.C.M of 2 and 4 is 4 and thus dividing this value by each denominator and multiplying the result by the corresponding numerators would produce

$$\frac{3x+1}{2} - \frac{5x-2}{4} = \frac{2(3x+1)-1(5x-2)}{4} = \frac{6x+2-5x+2}{4} = \frac{x+4}{4}$$

2.   Factorise $6xy+9y$.

The H.C.F of 6 and 9 is 3 and since y is present in each term we use $3y$ as our factor and thus the result is $3y(2x+3)$.

3.   If the universal set $U = \{1, 2, 3, 4, 5, 6, \ldots\ldots, 11, 12\}$ and
$A = \{factors\ of\ 12\}$, list the members of A and B.
$A = \{1,2,3,4,6,12\}$ and $B = \{2,3,5,7,11\}$

The solution is B since on factorising we get $2(3\underline{a} + 2\underline{b})$

Questions
Number Theory

Rosalind Atwell
Springer Memorial School, Barbados

Phillip Clarke
Parkinson Memorial School, Barbados

▸ Video solutions to these questions via the App "**CTS Maths**" for your smartphone or tablet. Details on http://CaribbeanTeachersSeries.com

QUESTIONS

[1] What is the smallest number that can be divided by 6, 8 and 9, leaving a remainder of 5?

[2] The first three (3) diagrams in a sequence are shown below.

n = 1   n = 2   n = 3

(a) Draw the FOURTH diagram in the sequence.

(b) The table below shows the number of squares in each diagram.

(c) Write down in terms of n the number of squares in the nth diagram of the sequence.

Diagram (n)	Formula	Number of Squares
1	4(1) - 3	1
2	4(2) - 3	5
3	4(3) - 3	9
10	x	y
z	------------	53

[3] In the number 3214_5 , determine

(a) the place value of the 3

(b) the total value of the 3

[4] Convert 101101_2

(a) to the decimal system (base 10)

(b) the base 7

[5] The figure below shows the first three diagrams in a sequence. Each diagram is made up of dots connected by line segments. In each diagram, there are d dots and l line segments.

(a) Draw the FOURTH diagram in the sequence.

(b) Complete the table by inserting the missing values at the rows marked (i) and (ii).

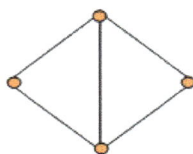
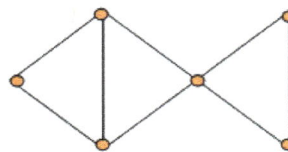

n = 1   n = 2   n = 3

n	No. of dots (d)	Formula for line segment (ℓ)	No. of line segments (ℓ)
1	3	3 + (1 − 1)	3
2	4	4 + (2 − 1)	5
3	6	6 + (3 − 1)	8
7	12	(i)	(ii)

[6] What is the largest number that can divide 18, 24 and 36 without leaving a remainder?

[7] If $2n - 3$ is an odd number, what is the odd number
(a) directly before this one.
(b) directly after this one.

[8] The following sets of numbers are defined as follows:

Natural Numbers (N) = {1, 2, 3, . . . }
Whole Numbers (W) = {0, 1, 2, . . . }
Integers (Z) = { . . . -4, -3, -2, -1, 0, 1, 2, 3, 4,...}
Rational Numbers (Q) = { p/q; p and q are integers; q ≠ 0}
Draw a Venn Diagram to show this information.

[9] A student uses sticks of different lengths to obtain polygons and she then connects all the vertices together. For example,

sticks = 3

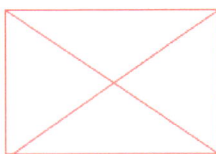

sticks = 6

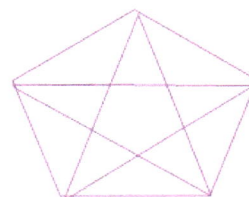

sticks = 10

How many sticks will she use in a hexagon?

[**10**] Find the square root of 729.

[**11**] Find the value of $\quad 3+9-6\div 2\times 8(3+1)$

[**12**] Simplify $\dfrac{-2q-3q\,(2q+5q)}{q}$.

[**13**] A sequence is given by 7, 11, 15, 19, 23
i.   Find the difference between each term.
ii.   Find the next two terms.
iii. Find the rule.
iv. Use the rule to find the 23rd and 30th values.

[**14**] Find the H.C.F of 15, 24, 36.

[**15**] Find the L.C.M of 15, 24, 36.

[**16**] Ms. Jones makes 63 cupcakes for her daughter's party. 27 are chocolate, 36 are coconut and she plans to place them in containers which will hold an equal type of cupcakes in each of them. Calculate:
i. The number of containers she should buy.
ii. The number of chocolate and coconut cupcakes in each container.

[**17**] Three traffic lights, X,Y and Z change their lights to green every 60, 90, and 105 seconds respectively. If they are both green at 11:54 p.m, calculate the time that they will all show green together again.

[**18**] Place the sets of Natural numbers **N**, Whole numbers **W**, and Rational numbers **Q** in the Venn diagram shown.

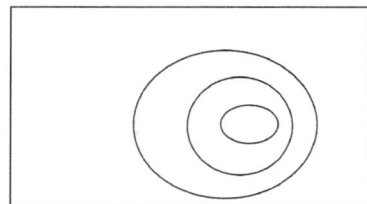

[**19**] For $x=-1$, $y=\dfrac{1}{2}$ and $z=4$ is $Q=(x+y)+z$ associative?

[**20**] If $A=\begin{pmatrix} a & b \\ c & d \end{pmatrix}$ and $B=\begin{pmatrix} p & q \\ r & s \end{pmatrix}$, prove that the commutative law is valid for only addition and not multiplication.

Chapter 2
Computation
Jerome Stuart
Graydon Sealy School

Barbados

Computation involves the use of mathematical operations to solve problems. There are four basic operations used to solve problems: addition, subtraction, multiplication and division. To calculate accurately we have to perform these operations in the correct order. If there is more than one operation in the question, the general rule is to perform the calculation in the brackets first, then divide or multiply, then add or subtract based on the operations given in the problem.

Rounding off numbers, adding, subtracting, multiplying and dividing fractions, converting from one unit to another – these are areas that normally present challenges. The intention of this chapter is to demonstrate how *easily* these calculations can be performed. This will greatly boost confidence in computational skills in and *out* of the classroom environment, for example, in everyday shopping and bill payments. Computation - the practical side of mathematics for real life situations!

Addition, subtraction, multiplication and division of real numbers

Addition and subtraction of fractions

- For common fractions, if the denominators are the same, add or subtract the numerators and place the result over the denominator.

- For common fractions, if the denominators are different, find the lowest common multiple (LCM) of the denominators, work out equivalent fractions and then add or subtract the fractions. If the result is an improper fraction, convert it to a mixed numeral.
- For mixed numbers, convert the fraction to an improper fraction and then follow the steps for I or II.

Multiplication of fractions

For common fractions, start by cancelling (divide by the highest common factor of a number in the numerator and a number in the denominator) then after cancelling work out the product of the numerators and place it over the product of the denominators.

Division of fractions

When dividing by a fraction, the first number remains unchanged except in a case with a mixed number that changes to an improper fraction then the division sign is changed to a multiplication sign, then you multiply by the reciprocal (inverse) of the fraction you were originally dividing by.

Examples

Using a calculator or otherwise determine the value of the following and give each answer as a fraction in its lowest terms.

1. $$\dfrac{\dfrac{2}{7} + \dfrac{5}{8}}{2\dfrac{1}{8}}$$

The LCM of 7 and 8 is 56. Notice that 16/56 is the equivalent fraction to 2/7.

Top $\dfrac{2}{7} + \dfrac{5}{8} = \dfrac{16}{56} + \dfrac{35}{56} = \dfrac{51}{56}$

Bottom $2\dfrac{1}{8} = \dfrac{8 \times 2 + 1}{8} = \dfrac{17}{8}$

$\dfrac{51}{56} \div \dfrac{17}{8} = \dfrac{51}{56} \times \dfrac{8}{17} = \dfrac{3}{7}$ in its lowest terms.

2. $$\frac{\frac{2}{7} \times \frac{5}{8}}{2\frac{1}{8}}$$

Top $\qquad \frac{2}{7} \times \frac{5}{8} = \frac{5}{28}$

Bottom $\qquad 2\frac{1}{8} = \frac{8 \times 2 + 1}{8} = \frac{17}{8}$

$\frac{5}{28} \div \frac{17}{8} = \frac{5}{28} \times \frac{8}{17} = \frac{10}{119}$ in its lowest terms.

3. $$\frac{\frac{2}{7} \times \frac{5}{8}}{2\frac{1}{8} - \frac{1}{4}}$$

Top $\qquad \frac{2}{7} \times \frac{5}{8} = \frac{5}{28}$

$2\frac{1}{8} - \frac{1}{4} = \frac{17}{8} - \frac{1}{4} = \frac{17}{8} - \frac{2}{8} = \frac{15}{8}$

Bottom $\frac{5}{28} \div \frac{15}{8} = \frac{5}{28} \times \frac{8}{15} = \frac{2}{21}$

4. $\left(\frac{3}{4}\right)^3 = \frac{3 \times 3 \times 3}{4 \times 4 \times 4} = \frac{27}{64}$

Division by decimals

When dividing by a decimal, we shift the decimal point to the right until we are dividing by a whole number. A shift takes place in the numerator that is consistent with the number of places moved to obtain the whole number in the denominator. A zero or zeros may be inserted in the numerator where needed to ensure that the number of places shifted in the numerator is the same as the number shifted in the denominator.

Example

Calculate the exact value of $\dfrac{1.7}{0.02} - 7.49^2$

Now $\dfrac{1.7}{0.02} = \dfrac{170}{2} = 87.5$

> We shifted the decimal point in the denominator two places to the right, so we shift the decimal point in the numerator two places to the right.

and $7.49^2 = 7.49 \times 7.49 = 56.1001$

> There are four digits to the right of the decimal point.

Thus 87.5 - 56.1001 = 31.3999 (exact value)

Multiplication of decimals

Decimals are multiplied in a similar way to whole numbers but the result must have the same number of digits to the right of the decimal point as the total number of digits to the right of the decimal point in the numbers being multiplied.

Example

Calculate the exact value of 4.25 x 3.93

```
    4.25
  x 3.93
    1275
   38250
  127500
  16.7025   Answer = 16.7025 (exact value)
```

Convert among fractions, percentages and decimals

- To convert a fraction to a decimal, divide the numerator by the denominator.
- To convert a fraction to a percentage, multiply by 100 and add the percent sign (%).

- To convert a decimal to a percentage, multiply by 100 and add the percent sign (%).
- To convert decimals to fractions multiply the decimal by the multiple of 10 that will make it whole, (this gives the numerator) then place the whole number over the same multiple of 10 (this gives the denominator). Then simplify to lowest terms.
- To convert a percentage to a fraction, place the percentage over 100 then cancel to lowest terms.
- To convert percentages to decimals divide the percentages by 100.

Examples

1.   Write $\dfrac{3}{8}$ as a decimal.

$8|3.^30^60^40$
$\quad 0.375$

Therefore $\dfrac{3}{8}$ = 0.375.

2.   Write $\dfrac{8}{40}$ as a percentage.

$\dfrac{8}{40}(100) \ = \ 20\%$

3.   Write 0.125 as a percentage.

$0.125(100) \ = \ 12.5\%$

4.   Write (a) 0.064 and (b) 7.25 as a fraction in its lowest terms.

(a) $0.064 = \dfrac{64}{1000} = \dfrac{8}{125}$. Therefore $0.064 = \dfrac{8}{125}$ (lowest terms)

(b) $7.25 = \dfrac{725}{100} = 7\dfrac{25}{100} = 7\dfrac{1}{4}$. Therefore $7.25 = 7\dfrac{1}{4}$

5.   Write (a) 6% and (b) 13% as a decimal.

(a) $100|6.^60^600$
$\qquad 0.06$

6% = 0.06

(b) 100|$1^13.^{13}0^{30}0$
         0.13

13% = 0.13

Conversion scales, metric scales, currency conversions and time

Key Facts

1 minute = 60 seconds	1 hour = 60 minutes	1 day = 24 hours
1 week = 7 days	1 year = 12 months	1 year = 365 days

- To convert from a smaller metric unit to a larger unit we *divide*.
- To convert from a larger metric unit to a smaller unit we *multiply*.
- To convert from 12 hour clock to the 24 hour clock we must ensure that it has four digits followed by the word *hours*.

  ‣ Midnight is 0000 hours on the 24 hour clock.
  ‣ Times between 12:00am and 1:00am put 00 followed by the 2 digits for the minutes of the time.
  ‣ For times from 1:00am to 9:59am, we place a zero in front of the time to convert it to the 24 hour time.
  ‣ For times from 10:00am to 12:59pm the digits remain the same
  ‣ For times from 1:00pm to 11:59pm add 12 hours to the time.

- To convert from the 24 hour clock to the 12 hour clock we generally write am for 'morning times' and pm for 'afternoon times'. In addition

  ‣ 24 hour times beginning with 00, begin with 12 on the 12 hour clock followed by a colon then the minutes then am.
  ‣ Remove the 0 at the beginning of the time for times from 0100 hours to 0959 hours and place a colon in between the hours and the minutes.
  ‣ For times from 1000 hours to 1259 hours keep the digits but separate the hours from the minutes with a colon.

▸ For afternoon times where the first two digits are larger than 12, subtract 12 hours then separate the hours from the minutes with a colon.

Examples

1.   Convert 37 cm to meters

$$37cm = \frac{37}{100}m = 0.37m$$

2.   Convert 1200 mm to meters

$$1200mm = \frac{1200}{1000}m = 1.2m$$

3.   Convert 1420 hours to 12 hour clock times.

$$14 - 12 = 2$$
$$1420 = 2:20pm$$

4.   Convert 1:55 am to 24 hour clock times.

$$1:55am = 0155 \; hours$$

5.   Convert 1:50 pm to 24 hour clock times.

$$1:50pm = 1350 \; hours$$

6.   Convert 10:40 am to 24 hour clock times.

$$10:40am = 1040 \; hours$$

7.   Convert 12:43 am to 24 hour clock times.

$$12:43am = 0043 \; hours$$

8.   Convert 42 months to years

$$42 \; months = \frac{42}{12} years = 3\frac{6}{12} = 3\frac{1}{2} years$$

Barbados   Bds $2.00
Jamaica   Jam $25.00
East Caribbean   EC $3.30
Trinidad   TT $6.00

Example

A Jamaican man changed Jam $32000 into the United States currency. A 1% tax was charged on the total foreign exchange transaction. He spent US$780, how much US$ did he return with to Jamaica?

For the table, we see that US $1 = Jam $25

$$Jam\ \$32000\ =\ US\$\frac{32000}{25} = US\ \$1280$$

$$1\%\ tax\ =\frac{1}{100}\ US\$1280\ =\ US\ \$128$$

He received US $(1280 - 128) = US $1152

After spending US $780, he returned with US $1152 - $780 = US $372

The rules for rounding are as follows:

• Count one more digit than is required for the number of decimal places or significant figures stated in the question, if this digit is a 5, 6, 7, 8 or 9 add 1 to the digit that comes to the left of this digit and drop all digits to the right of this digit. If the digit is a 0, 1, 2, 3, or 4 keep all the digits to the left the same and drop the digits to the right.

• When expressing a value to a certain number of decimal places, begin to count from the first digit to the right of the decimal point then follow the rules for rounding above.

• When expressing a value to a certain number of significant figures, begin to count the digits in the value from the first non-zero digit on the left then follow the rules for rounding above. In the case of whole numbers a zero or zeros are placed after rounding to ensure that the number does not lose its significance. (See example 1)

Examples

Using a calculator or otherwise determine the value of

1.   62 x 305 and write the answer exactly and also significant to 1, 2 and 3 figures.

$62(305) = 18910$ exactly
$62(305) = 20000$ correct to 1 significant figure
$62(305) = 19000$ correct to 2 significant figures
$62(305) = 18900$ correct to 3 significant figures

2. $0.146072 \div 6.2$ and write the answer exactly and also significant to 1, 2 and 3 figures.
$0.146072 \div 6.2 = 0.02356$ exactly
$0.146072 \div 6.2 = 0.02$ correct to 1 significant figure
$0.146072 \div 6.2 = 0.024$ correct to 2 significant figures
$0.146072 \div 6.2 = 0.0236$ correct to 3 significant figures

3. $0.146072 \div 6.2$ and write the answer exactly and also to decimal 1, 2 and 3 places.

$0.146072 \div 6.2 = 0.02356$ exactly
$0.146072 \div 6.2 = 0.0$ correct to 1 decimal value (1 d.p)
$0.146072 \div 6.2 = 0.02$ correct to 2 d.p
$0.146072 \div 6.2 = 0.024$ correct to 3 d.p

4.   5.326 x 4.9 and write the answer exactly and also to decimal 1, 2 and 3 places.

$5.326(4.9) = 26.0974$ exactly
$5.326(4.9) = 26.1$ correct to 1 decimal place
$5.326(4.9) = 26.10$ correct to 2 decimal place
$5.326(4.9) = 26.097$ correct to 3 decimal places

Writing rational numbers in standard form

A number is in standard form if it is in the form (A x 10^n) where A is a number greater than or equal to 1 but less than 10 and n is an integer. Standard form is used to represent very large or very small numbers.

Examples

Write the following numbers in standard form.

1. $3045.6 = 3.0456 \times 1000 = 3.0456 \times 10^3$

2. $942 = 9.42 \times 100 = 9.42 \times 10^2$

3. $0.00356 = 3.56 \times \dfrac{1}{1000} = 3.56 \times 10^{-3}$

Calculating fractions or percentages of quantities

- To calculate a fraction of a quantity, write the fraction followed by a multiplication sign then the quantity; cancel those values that can be cancelled then multiply the two numerators by each other and the denominators by each other.

- To calculate a percentage of a quantity write the percentage over 100 followed by a multiplication sign then the quantity; cancel those values that can be cancelled then multiply the two numerators by each other and the denominators by each other.

Examples

1.   Calculate 45% of 350

$$\frac{45}{100}\left(\frac{350}{1}\right) = 157.5$$

2.   Calculate $\dfrac{7}{12}$ of $360

$$\frac{7}{12}(\$360) = \$210$$

Express one quantity as a fraction or percentage of another

To express a quantity as a fraction, place the quantity over the total then multiply by $\dfrac{100}{1}$ then place a % sign.

Examples

1.   36 out of 200 bulbs were faculty, what percentage of the bulbs were faulty?

$$\frac{36}{200} \times \frac{100}{1}\% = 18\%$$

Thus, 18% of the bulbs were faculty.

2.   350 out of 600 children were girls, what fraction were girls?

$$\frac{350}{600} = \frac{7}{12}$$ of the children were girls.

Compare two quantities using ratios

To compare two quantities using ratios, first make sure that they share the same units otherwise convert one of them to the unit of the other. It is also useful to reduce both of them by cancelling then separate the results obtained by a colon.

Examples

1.   State 8m to 50cm as a ratio in its lowest terms.

8m = 800cm
800 : 50 = 16 : 1 in lowest terms

> Ratios should always be reduced to lowest terms.

2.   Express $6.00 : $0.75 as a ratio in its lowest terms.

$6: 75 cents
$6 = 600 cents
600 : 75 = 8 : 1 in lowest terms

> In comparing ratios, compare ratios with the same units.

Divide a quantity in a given ratio

When dividing quantities in ratios, first sum the ratios, then calculate the fraction allotted to each part.

Examples

1.   A piece of rope is 120cm is divided into three pieces in the ratio of 5:3:2. Calculate the length of the longest and shortest piece.

Since $5+3+2=10$

The length of the longest piece = $\frac{5}{10}$ x 120 = $60cm$

The length of the shortest piece = $\frac{2}{10}$ x 120 = $24cm$

2.   There are 250 flags in a bag. The colours are white, green, red and black in the ratio of 3:2:4:1. How much of each colour flag is there in the bag?

Since $3+2+4+1 = 10$

$\frac{3}{10}$ x 250 = 75   (White)

$\frac{2}{10}$ x 250 = 150   (Green)

$\frac{4}{10}$ x 250 = 100   (Red)

$\frac{1}{10}$ x 250 = 25   (Black)

Fractions, decimals, percentages, ratios, rates and proportions, arithmetic mean

Arithmetic mean is calculated for a list of numbers by adding the numbers given in the list and dividing them by how many numbers were given. Direct proportion occurs when one quantity increases as another increases and inverse proportion occurs when one quantity decreases as another increases.

Examples

1.   A woman distributes money to her three children in the ratio 4:2:5. If the smallest share is $3.00 what was the total amount of money shared?

2 proportional parts = $3.00
1 proportional parts = $1.50
Total number of proportional parts = $4+2+5 = 11$
The total amount of money shared = $1.50 x 11 = $16.50

2.   A truck travels 320 km on 8 litres of diesel. How many litres will it take to complete a trip of 480 km?

Amount of litres for 320 km = 8l

Then the amount of litres for 1 km = $\dfrac{8}{320}$

The amount of litres for $480km = \dfrac{8}{320}$ x $480 = 12l$

Therefore 12 l are needed for a trip of 480 km.

3.   A pasture feeds 15 cows for 8 days. How many days would the same pasture feed 6 cows?

The days of feeding for 15 cows = 8 days
The days of feeding for 1 cow = 15 x 8 = 120 days

So the pasture can feed 6 cows in $\dfrac{120}{6} = 20$ days

4.   Alex's mean score for 12 tests is 64. What was the total score for all twelve tests?

Alex's total for the 12 tests is 12 x 64 = 768.

5.   Tracey's marks in 5 tests were 67, 84, 56, 94 and 73.

a)   What was her mean score for the first 5 tests?

Total for the 5 tests = 67 + 84 + 56 + 94 + 73 = 374

Mean score = $\dfrac{374}{5} = 74.8$

b)   What did she score in the 6th test if the mean for the 6 marks were 68?

Mean of six numbers is 68, so the total = 68 x 6 = 456
Tracey therefore scored (456 - 374) = 82 in the sixth test.

6.   Express 34cm as a percentage of 2m.

$\dfrac{34}{200}$ x 100% = 17%

First convert to the same units.

7.   Samuel's property has a rateable value of $7,400. How much will the rates be at 35 cents in the $1?

Rate payable = Rateable value x Rate in dollars
= $7400 x $0.35
= $2590

8.   In St. Lucia, Mr Grandson pays $420 on his property which has a rateable value of $3000. What is the rate in the dollar?

The rate on $3000 is $420

The rate on $1 is $\dfrac{420}{3000}$ = 0.14 cents

The rate is 14 cents in the dollar.

Property Rates = Rateable Value (RV) x the rate in the dollar

Questions
Computation

Stuart Mayers
Combermere School
Barbados

Jerome Stuart
Graydon Sealy School
Barbados

QUESTIONS

[**1**] Written as a single fraction, is $\frac{1}{2} - \frac{1}{4} \times \frac{5}{6}$

   A.   0

   B. $\frac{5}{24}$

   C. $\frac{7}{24}$

   D. $\frac{15}{24}$

> ‣ Video solutions to these questions via the App "**CTS Maths**" for your smartphone or tablet. Details on http://CaribbeanTeachersSeries.com

[**2**] What is 0.000376 in standard form?
  A. 37.6×10^{-5}
  B. 3.76×10^{-4}
  C. 0.376×10^{-3}
  D. 3.76×10^{4}

[**3**] James completed $\frac{1}{3}$ of a job. Henry completed $\frac{1}{4}$ of the remainder. How much of the job is still left to be done?

[**4**] Calculate the exact value of $\sqrt{6}\,\frac{1}{4} + \sqrt{20}\,\frac{1}{4}$

[**5**] The value of 2783 to 2 significant figures is
  A.   27
  B.   28
  C.   2700
  D.   2800

[**6**] A sum of money was shared among Mary, Jane and Sue in the ratio of 2:3:5. Jane got $270. Find the sum of money shared.

[**7**] The scale of a map is 1:15000. If the distance between two points on the map is 9 cm, calculate the actual distance between the two points. Give your answer in kilometers.

[**8**] Using the table of exchange rates below, convert US $950 into TT dollars.

US $	Trinidad & Tobago (TT) $	Barbados (BDS) $
1	5.97	1.98

[**9**] Using the table of exchange rates below, convert TT $650 to Barbados dollars. Give your answer to the nearest cent.

US $	Trinidad & Tobago (TT) $	Barbados (BDS) $
1	5.97	1.98

[**10**] What percentage of 30 is 12?
  A.   18
  B.   40
  C.   42
  D.   250

[**11**] A man leaves an inheritance of $350,000 to his three sons Charis, David and Joseph in the ratio of 7:4:3. How much more money did Charis receive than David?

[**12**] What is the total number of minutes in 5 days, $3\frac{1}{4}$ hours?

[**13**] A plane departs from Singapore at 1525 hours and arrives at its destination the next day at 0114 hours. How long was the flight?

[**14**] Calculate $\sqrt{2.364 \times 0.4}$ giving your answer to 3 decimal places.

[**15**] Write 5632 in standard form.

[**16**] Calculate $4\frac{2}{7} - 1\frac{3}{4} \times \frac{6}{7}$ giving your answer as a fraction.

[**17**] Calculate the exact value of 12.44 x 5.88/1.4.

[**18**] If 10 men can tile a house in 21 days. How long will it take 15 men to tile it assuming that they all work at the same rate?

[**19**] Write 0.02567 correct to (a) 1 significant figure (b) 2 significant figures.

[**20**] A typist is required to type a manuscript with 30,000 words. If her accuracy is 97%, how many errors did she make?

Chapter 3
Consumer Arithmetic
Tameisha Cozier
Alleyne School
Barbados

Congratulations on being one of the millions of students who have reached this level and are determined to go even further into investigating the mathematics of the world around us. In this chapter, we will be looking at consumer arithmetic. "What is it?" you may wonder. You are certain to have heard the words before. Simply put, Consumer Arithmetic is the branch of mathematics which deals with real world problems. You will investigate and solve problems such as discounts, profit and loss, sales tax, and utility bills, just to name a few.

Good luck students!

Covered in this Chapter

- Profit, Sales Tax & Discount
- Loss and Percentage Loss
- Exchange Rate
- Simple Interest
- Compound Interest
- Depreciation
- Hire Purchase
- Utility Bills
- Mortgages

Profit, Sales Tax & Discount

Profit implies the **selling price** is **greater** than the **cost price**. It can be found using the formula:

Selling Price - Cost price

Discount is a reduction in the cost price of an item, which is usually given as a percentage. Sales Tax is a fixed percentage added to the cost of an item(s).

Example

A department store bought 50 backpacks for $1500. Each is sold making a profit of 35% on each of them.

(a) How much does a customer pay for one (1) backpack?

Cost of 1 backpack $= \dfrac{1500}{50} =$ $30.00. With 35% profit, the Selling Price $= \dfrac{135}{100}$ x 30
= $40. 50 is the amount customer pays.

(b) V.A.T is added on at 17.5%. What is the actual price a customer pays for one (1) backpack?

Since the V.A.T is 17.5 %, the cost to customer $= \dfrac{117.5}{100}$ x $40.50 = $47.59 to the

nearest cent.

(c) A discount of 5% is given on the selling price V.A.T inclusive. What is the price a customer would now pay for a backpack?

Discount $= \dfrac{5}{100}$ x $47.59 = $2.38 (to the nearest cent)
Customer pays: $47.59 - $2.38 = $45.21

> Inclusive means the tax is included in the selling price.

Percentage Loss

Loss is the mathematical opposite of the Profit. If a loss is made, the **cost price** is **greater** than the **selling price**.   Use the formula:

Cost price - Selling Price

Example

BLUEBOX Creative purchased some laptops for resale at $795 each. However, during transport, some laptops were damaged externally. The cost to a customer is now $465. Calculate the percentage loss to BLUEBOX Creative.

Loss   = Cost Price – Selling Price
       = $795 - $465
       = $330

Percentage Loss = $\dfrac{Loss}{CostPrice}$ x 100%

Therefore, the Percentage Loss to BLUEBOX Creative

$= \dfrac{330}{795}$ x 100
= 41.509 %
= 42% to the nearest whole number.

Exchange Rate

An exchange rate is required to change from one currency to another using specific rates.

Example

A tourist leaving St. Kitts for Trinidad converts EC $6600 into US travellers cheques. On arrival in Trinidad he converts the US dollars to TT dollars and the bank charges amount to TT $92.30 on the total transaction.

Calculate the amount the trader received in
(a) US dollars

RATES
US $ 1 = EC $ 2.75
US $ 1 = TT $ 6.75

EC $2.75 = US $1

Therefore, EC $ 1 = US $ $\dfrac{1}{2.75}$

EC $6600 = US $ $\dfrac{1}{2.75}$ x $6600 = US $2400 the trader received

(b) TT dollars after charges

> When dealing with money, your answers must be written to two decimal places (2 d.p.) so be sure that you know how to round off numbers.

US $1 = TT $6.75
Therefore US $2400 = TT $6.75 x 2400 = TT $16,200
After charges = TT $16,200 – TT $92.30 = $16,107.70

Simple Interest

As the name suggests, this is an easy method of calculating the interest charge on a loan. To calculate the simple interest we use the formula $I = \dfrac{PRT}{100}$, where

  I = Interest
  P = Principal
  R = Rate percent
  T = Time in **years**

Examples

1.   Calculate the simple interest earned on $4500 invested into a Financial Company for 5 years at 8% per annum.

P = $4500, R = 8% and T = 5 years

Therefore $I = \dfrac{4500(8)(5)}{100}$ = $1800

> But what if I want to find out T? Simply rearrange the formula to make T the subject i.e. $T = (100)I/PR$

2.   Determine the length of time that $550 will be the simple interest on $100 which was invested at 11% per annum.

I = $550, P = $100, R = 11%

Therefore $T = \dfrac{100I}{PR} = \dfrac{100(550)}{100(11)} = 5$ years

Similarly, $R = \dfrac{100I}{PT}$ and $P = \dfrac{100I}{RT}$.

Try these questions below on your own.

(a) Mrs. Lange invested $10, 500 into Percy Insurance Ltd. for a period of 6 years. If the interest accumulated amounted to $4750, calculate the rate percent. (Give your answer to 1 d.p) (**Ans** 7.5%)

(b) How much money should a person invest in a company in order to collect $11550 interest for 7 years at 8.25 % per annum? (**Ans** $20,000)

Compound Interest

The **Compound Interest** is interest calculated on the initial **principal** and also on the accumulated or accruing **interest** in the years following. For example, over two years, the Principal P_2 at the beginning of the second year is given by $P_2 = P_1 + I_1$, where Principal P_1 is amount invested at the beginning of the first year and I_1 is the interest for the first year. The total compound interest is then $I_1 + I_2$, where
$I_1 = \dfrac{P_1 R(1)}{100}$ and $I_2 = \dfrac{P_2 R(1)}{100}$.

The amount accruing can be calculated using the formula $A = P\left(1 + \dfrac{R}{100}\right)^n$

where

        A = Amount
        P = Principal
        R = Rate percent
        n = Number of years

> The time is one year in each case.

Examples

1.   Calculate the compound interest earned if Tammy Chase has invested $1500 for two years at 4% per annum into a bank.

▸ This can be done using the same formula for Simple Interest and adding on (compounding) the interest for subsequent years.

Interest for year one $I_1 = \dfrac{P_1 RT}{100} = \dfrac{1500(4)(1)}{100} = \$60$

Only for the first year!

Principal P_2 at the beginning of the second year is given by

$P_2 = P_1 + I_1$
$= \$1500 + \$60$
$= \$1560$

For the second year

Hence $I_2 = \dfrac{P_2 RT}{100} = \dfrac{1560(4)(1)}{100} = \$62.40$

Finally, the total compound interest = $I_1 + I_2$ = \$60 + \$62.40 = \$122.40

2.   Mr. Stuart invested \$25000 into BLUEBOX Financial for a period of 3 years. Calculate:

(a) The amount accruing at 7.5% compound interest per annum.

$A = P\left(1 + \dfrac{R}{100}\right)^n$

$= \$25000\left(1 + \dfrac{7.5}{100}\right)^3$

$= \$25000(1 + 0.075)^3$

$= \$31057.50$

(b) The compound interest earned in the 3 years.

Compound Interest (C.I) = Amount – Principal
= \$31057.50 - \$25000.00
= \$6057.50

Depreciation

When an item loses its original value and costs less than it was bought for, it is said to have depreciated. Examples of items which depreciate in value are cars, electronics and toys, just to name a few.

Example

Chelsea Garner bought a Mazda 3 for $48,000 in 2011. How much money can she expect to receive for her car if she decides to sell 4 years later in 2015, if the rate of depreciation is 7.5% per year?

Depreciation at end of the 1ˢᵗ year $= \dfrac{7.5}{100}(\$48000) = \$3600$

Value of car after 1ˢᵗ year $= \$48000 - \$3600 = \$44400$

Depreciation at end of the 2ⁿᵈ year $= \dfrac{7.5}{100}(\$44400) = \$3330$

Value of car after 2ⁿᵈ year $= \$44400 - \$3330 = \$41070$

> $37989.75 is the value at the beginning of the fourth year.

Depreciation at end of the 3ʳᵈ year $= \dfrac{7.5}{100}(\$41070) = \$3080.25$

Value of car after 3ʳᵈ year $= \$41070 - \$3080.25 = \$37989.75$

Depreciation at end of the 4ᵗʰ year $= \dfrac{7.5}{100}(\$37989.75) = \$2849.23$ (correct to 2 d.p)

Value of car after 4ᵗʰ year $= \$37989.75 - \$2849.23 = \$35140.52$

You may be thinking to yourself, "How can I do all of that working in an exam?" That's easy, use the formula:

$$A = P\left(1 - \frac{R}{100}\right)^{n}$$

where

A = Value of item after n years
P = Original cost of item
R = Rate percent
n = Number of years under consideration

For the same example above, $A = 48000\left(1 - \dfrac{7.5}{100}\right)^{4} = \$35140.52$

Hire Purchase

Hire purchase is the method of payment used by businesses when persons do not have all of the cash available to pay for an item. For example, items such as TVs, furniture and appliances can be bought on hire purchase.

> Hire Purchase results in a higher payment.

Example

Hans Johnson wishes to purchase a living room suite valued at $2199.90. However, not being able to afford it at that price, he decides buy it on hire purchase based on the following terms:

- A deposit of 10% must be paid.
- In addition to 18 monthly instalments of $165.00

(a) Calculate the cost of the suite on hire purchase.

The deposit is given by $\dfrac{10}{100}(\$2199.90) = \$219.99$

Total monthly instalments = $165.00 x 18 = $ 2970.00

Hire Purchase Price = Deposit + Monthly Instalments
$$= \$219.99 + \$2970.00$$
$$= \$3189.99$$

(b) What is the difference between the hire purchase price and the cash price?

Difference = Hire Purchase Price – Cash Price
$$= \$3189.99 - \$2199.90 = \$990.09$$

(c) Express this difference as a percentage of the cash price.

Percentage Difference = $\dfrac{\$990.09}{\$2199.90}$ x 100 = 45%

Utility Bills

Another reason for understanding Consumer Arithmetic is in the calculation of utility bills. What are utilities? These are the services that are provided to the public such as electricity and water. Let's look at an example of how to calculate an electricity bill.

Example

Look at a sample of Charlene's electricity bill below and use the information given to determine her total electric bill.

Account No: 112233-4455	Charlene Calor	Address
Date: April 30th 2015		BlueBox Drive, Barbados
Previous Reading: 5596 KWH	Present Reading: 6114 KWH	
Rates: First 100 KHW = $ 0.025 Next 250 KHW = $ 0.05 Remainder = $ 0.10 Meter Rental = $4.35		

(a) Calculate the number of KWH Charlene used in the month of April.
Energy used = Present Reading – Previous Reading = 6114 – 5596 = 518 KWH

(b) Calculate the total energy cost of Charlene's bill.

First 100 kWh = 100 x 0.025 = $2.50
Next 250 kWh = 250 x 0.050 = $12.50
Remainder = 518 – (100 + 250) = 518 – 350 = 168 kWh
Cost   = 168 x 0.10 = $16.80
Energy Cost = $2.50 + $12.50 + $16.80 = $31.80

(c) Calculate the total cost of Charlene's electricity bill if there is 17.5 % V.A.T added on to it.

Electrical Cost = Energy cost + Metre Rental
$\qquad$ = \$31. 80 + \$4.35 = \$36.15

V.A.T. = $\dfrac{17.5}{100}$ x \$36.15 = \$ 6.33 (to 2 d.p)

Therefore Charlene's total bill = \$36.15 + \$ 6.33 = \$ 42.48

Mortgage

This is a legal agreement between an individual or persons wishing to buy property and do not have cash readily available. The individuals involved agree to pay a fixed sum of money to the lending institution over a period of time.

Example

Jeremy and Keona Taylor purchased a house in 2010 for \$220,000. They paid a deposit of 10% by cheque and received a mortgage for the remainder. Calculate,

(a) The amount of the mortgage

Deposit = $\dfrac{10}{100}$ x \$ 220,000 = \$ 22,000

> 1 year = 12 months so multiply the monthly payments by 12

Mortgage= \$ 220,000 - \$22,000 = \$ 198,000

(b) The total repaid if monthly instalments of \$1135.15 for 25 years are to be paid.
Mortgage repayments = 25 x 12 x \$1135.15 = \$ 340,545.00

> Well dear students, you have come to the end of another chapter. I hope that now you are more comfortable tackling consumer arithmetic questions. They are very simple once you remember the basics and the formulae. I wish you all the success you deserve. Keep practising!

Questions
Consumer Arithmetic
Tameisha Cozier
Alleyne School
Barbados

▸ Video solutions to these questions via the App **"CTS Maths"** for your smartphone or tablet. Details on http://CaribbeanTeachersSeries.com

QUESTIONS

[1] Ravi, a construction worker, works a basic week of 40 hours at a rate of $12.05 per hour.
(a) Calculate Ravi's basic wage for one week.
(b) On one Saturday, he worked 7 hours overtime at time-and-a-half. Determine Ravi's gross wage for that week.

[2] The table below shows the items Amanda purchased on a shopping trip to Sheraton Mall. Calculate the missing values: A, B, C, D and E.

Item	Quantity	Unit Price($)	Total Price($)	
Jeans	2	$95.00	A	
Career Shirts	3	B	$96.15	
Casual Shirts	C	$15.04	$75.20	Pre-V.A.T total $366.35
V.A.T (17.5%)	D			Overall total ($) E

[**3**] Mr. Yearwood is a bachelor who earns $5035.28 monthly.
(a) Calculate his gross annual salary.
(b) How much money does he pay yearly in income tax if it is calculated at 12% of his gross annual salary?
(c) Determine Mr. Yearwood's net annual salary.

[**4**] Sarah borrowed $3050 from Stoutes Bank at 7.25% simple interest per annum for 3 years. Determine,
(a) the simple interest to be paid.
(b) the total amount of money to be repaid.

[**5**] Kelly is paid $225 in simple interest at 3% per annum for 4 years. How much money did she invest?

[**6**] A trader leaving Antigua for Trinidad converts EC $3300 into US travelers cheques. On arrival in Trinidad he converts the US dollars to TT dollars and the bank charges amount to TT $6.75. Calculate the amount the trader received
(a) in US dollars.
(b) in TT dollars after charges.

RATES: US $1 = EC $2.75 and US $1 = TT $6.75

[**7**] BLUEBOX Financial has a special on mortgages where successful applicants can obtain an 85% mortgage for a 25 year period. Sam and Eva wish to purchase a house valued at $395,000. Calculate,
(a) the deposit.
(b) the mortgage.
(c) the total amount to be repaid to BLUEBOX Financial if monthly payments are $2175.
(d) the interest to be repaid.
(e) the total amount paid for the house.

[**8**] A salesman bought 200 pairs of jeans for $15000. He sells them making a profit of 20%.
(a) How much does a customer pay for 1 pair of jeans?
(b) V.A.T is added on at 17.5%. What is the price a customer pays for one pair of jeans?
(c) A discount of 15% is given on the selling price, V.A.T inclusive. What is the price a customer would now pay for a pair of jeans?

[9] Mrs. Batch owns a bakery. A customer wishes to purchase 50 cupcakes at $1.75 each and 75 brownies at $2.10 each. Calculate,
(a) the total cost of the cupcakes and brownies to a customer.
(b) the cost a customer actually pays if a sales tax of 15% is added on for all sales.

[10] BLUEBOX Creative purchased some laser printers for resale at $405 each. However, some printers were damaged during transport. The cost to a customer for a damaged printer is now $265. Calculate the percentage loss.

[11] Determine the length of time that $550 will be the Simple Interest on $100 which was invested at 11% annum.

[12] Destiny invested $700 in BLUEBOX Financial for a period of 5 years. The interest accumulated was $210. What is the rate percent?

[13] A BLUEPAD car valued at $60000 in 2011 depreciates in value by 12% per year. What is the value in 2013.

[14] BLUEBOX Lifestyles has the Perfect living room suite for $1475.00 cash. A customer can obtain this suite on hire purchase terms by paying a deposit of 15% and the remainder at $66.25 for 24 months. Calculate, (a) the deposit. (b) the total monthly instalments. (c) the total hire purchase price. (d) the difference between the hire purchase price and cost price as a percentage of the cash price.

[15] The electricity bill for Alaina Branch for October 2012 is seen in the table below. Energy and fuel charges are calculated based on the number of units (KWh) used. Calculate the values indicated by the letters A, B, C, D and E.

	Present Reading	Previous Reading	No. of KWh Used
	9946	9684	A
		Rate per KWh	
CHARGES	Service Charge		$10.00
	Energy Charge	0.19	B
	Fuel Charge	C	$117.90
Sub-Total			D
VAT @ 15%			E
Current Total			$204.33

[**16**] Study the table below which shows the water rate of a particular island.

First 1000 gallons	$75.00	Cost per gallon A
Next 1500 gallons	B	Cost per gallon $0.04
Fixed Charge	$32.00	

(a) Determine the values A and B in your table.
(b) Find the number of gallons of water used for one month if the remaining units are charged at a rate of $0.05 and the total water bill is $231.25.

[**17**] A monthly telephone bill is made up of a $25 rental charge, while calls cost 40 cents each. How many calls were made if the monthly bill is $110.60?

[**18**] BLUEBOX Financial is offering an investment deal which pays simple interest at 9% per annum for a savings bond. A competitor is offering a savings account deal which pays compound interest at the same rate.

(a) If Carl wishes to invest $10,000 for 2 years, which company should he choose?

(b) Calculate to the nearest cent, the difference between the amounts of the two investments at the end of 2 years.

[**19**] Calculate the amount of simple interest you would earn after 6 months if you invested $4575 into a financial institution which has a rate of 4.5% for such investments.

[**20**] Mr. Drakes has a decision to make. He has $3500 to invest. Bank A offers 5.5% compound interest for 5 years and Bank B offers 6% interest for 3 years. Use the compound interest formula to calculate which bank will give him the better compound interest.

Chapter 4
Algebra : I
Corey Worrell
Lester Vaughan School
Barbados

Algebra is a part of the Mathematics family, which utilizes symbols and letters in conjunction with arithmetic to represent or solve expressions, equations and problems.

History of Algebra
The word algebra was derived from the arabic word *al-jabr* and is translated as restoration or completion.

Why Should I learn Algebra?
Components of Algebra are found in most aspects of mathematics and as a result is very useful in helping to find solutions to problems. Additionally, Algebra is very abstract and this forces the brain to think in new patterns, which encourages brain development.

Why the use of letters?
In algebra, symbols are used to represent unknown numbers or quantities. The general practice globally, is to use any letter from the english alphabet to represent these unknowns but the most commonly used letters are x, y, z, a, b and c.

Common terms used in algebra and their meaning

- A *Variable* is any letter that is used to represent an unknown.

> Knowing the terms will help guide you towards your solutions.

- A *Constant* is a number by itself in an algebraic expression.

- A *Coefficient* is a number used to multiply a variable.

- An *Algebraic Term* is the product of variables and constants.

- An *Algebraic Expression* is an expression made up of constants and variables within arithmetic operations.

- *Equation* is an algebraic expression that is equal to a number or another expression.

- **Linear** Expression or Equation has 1 as the highest power (exponent) of the variable $4y - 6 = 2$ which can be re-written as $4y^1 - 6 = 2$

- **Quadratic** Expression or Equation has 2 as the highest power (exponent) of the variable. For example $8x^2 + 16 = 4$

$$3y + 5 = 16$$

Translate a verbal phrase into an algebraic statement

Read the verbal statement given and translate it into an equation. Thus, "I choose a number, add 12 to it and the result is 18" is translated into $x + 12 = 18$ where x is the unknown number.

Examples

> Always state the symbol representing the unknown

1. "I choose a number, multiply it by 6 and the result is 24"

Let y = unknown number, then $6y = 24$

2. "Tommy has four times as many shirts as Mark and in total they have 60 shirts"

Let x = number of shirts Mark has, therefore Tommy has $4x + x = 60$

Translate an algebraic statement into a verbal phrase

Read the equation given and translate it into a sentence.   For example, x + 4 = 9 would be translated into the sentence "when I add four to a certain number, the results is nine".

Example

5t -7 = 2   "when seven is taken from five times a certain number, the result is two".

Arithmetic operations

Below are examples of arithmetic operations, which make use of **directed numbers**, which are numbers that can be negative as well as positive.

Examples

1. 4 - 6 = -2
2. 3 + 16 = 19
3. 7 - 8 = -1
4. -10 - 3 = -13
5.   -6 x 4 = -24
6.   -5 x -7 = 35
7. 10 / (-2) = -5
8. -3 (-6 + 6) = 0

Simplify an algebraic expression (simplification)

To simplify an algebraic expression, reduce the given expression to a simpler form.

Examples

1. x + x = 2x
2. 6y + 2y = 8y
3. 12t - 2t = 10t
4. 16k - 21k = -5k
5. 2d - 3f + 5d - 9f = 7d - 12f
6. a x a = a^2
7. $3a^4b \times 2a^3b^4 = 6a^7b^5$

Only add or subtract like or similar terms.

8. $x^5y / x^2y = x^3$
9. $14a^3 / 7a^8y^2 = 2 / a^5y^2 = 2a^{-5}y^{-2}$

Solve an algebraic expression (substitution)

To solve an algebraic expression, substitute the value of the variables into the given equation.

Examples

If x = 3, y = 2 and z = 1, solve the following algebraic expressions

1. $x + y = 3 + 2 = 5$
2. $3y + z = 3(2) + 1 = (3x2) + 1 = 6 + 1 = 7$
3. $4z + 2x = 4(1) + 2(3) = (4x1) + (2x3) = 4 + 6 = 10$
4. $-10y + 2z = -10(2) + 2(1) = -20 + 2 = -18$
5. $x^2 = 3^2 = 9$
6. $4z^3 = 4(1^3) = 4(1) = 4$
7. $5x^2 - 2y^3 = 5(3^2) - 2(2^3) = 5(9) - 2(8) = 45 - 16 = 29$
8. $3xyz = 3(3)(2)(1) = 3x3x2x1 = 18$

> Substitute can be interpreted as exchange or replace.

Binary operations

A binary operation is an operation or rule used to combine two objects of a given type to obtain a result of a similar type.

In primary school, you used binary operations. The objects used were whole numbers and the binary operation being examined was multiplication, division, addition and subtraction.

Example

Given that a * b represents 2a + 3b, determine the value of the following.

1. 1 * 3

The first variable is 'a' and the second variable is 'b', therefore a=1 and b=3. Hence, 1 * 3 = 2(1) + 3(3) = 2 + 9 = 11

2. $4 * -2 = 2(4) + 3(-2) = 8 - 6 = 2$

3. $5 * (2 * 3) = 5 * [2(2) + 3(3)] = 5 * [4 + 9] = 5 * 13 = 2(5) + 3(13) = 25 + 39 = 64$

Expand an algebraic expression using distributive Law (distribution)

Expand the brackets using the distributive law and simplify.

Examples

1. $3(x + 4) = 3x + 12$
2. $6(1 + a) = 6 + 6a$
3. $5(y - 4) = 5y - 20$
4. $2(a + 3) + 5(a + 4) = 2a + 6 + 5a + 20 = 7a + 26$
5. $4x(x - 2) = 4x^2 - 8$
6. $7y^2(y^3 + 2z) = 7y^5 + 14y^2z$
7. $-(x + 2) = -x - 2$
8. $-5(t - 5) = -5t + 25$
9. $-2(-3r + 4) = 6r - 8$
10. $-4(a + 3) + 2(a + 8) = -4a - 12 + 2a + 16 = -2a + 4$
11. $5 - 2(3x - 4) = 5 - 6x + 8 = -6x + 13$

Things to Remember
- Always use brackets
- Group the like terms
- Apply the laws of indices where applicable

Solving quadratic equations (perfect squares)

In order to solve a quadratic equation, you must first know how to factorize it. For example, consider the equation $x^2 - 14x + 49 = 0$

$x^2 - 14x + 49 = 0$
$(x - 7)(x - 7) = 0$
$(x - 7) = 0$
$x = 7$

$(a + b)^2 = a^2 + 2ab + b^2$

$(a - b)^2 = a^2 - 2ab + b^2$

Here is another example.

$x^2 + 10x + 25 = 0$
$(x + 5)(x + 5) = 0$
$(x + 5) = 0$
$x = -5$

Questions

Solve the following quadratic equations.

a) $x^2 + 4x + 4 = 0$

$(x+2)(x-2)$ therefore $x = 2$ and $x = -2$

b) $x^2 + 6x + 9 = 0$
$(x+3)(x+3)$ therefore $x = -3$

c) $x^2 + 18x + 81 = 0$

$(x+9)(x+9)$ therefore $x = -9$

d) $16x^2 + 8x + 1 = 0$

$(4x+1)(4x+1)$ therefore $x = -\dfrac{1}{4}$

e) $81x^2 - 36x + 4 = 0$

$(9x+2)(9x+2)$ therefore $x = -\dfrac{2}{4}$

Solving quadratic equations (difference of two squares)

It is important to know square values such as $2^2 = 4$, $6^2 = 36$, $9^2 = 81$ etc. For example, consider the equation $x^2 - 4 = 0$

$x^2 - 4 = 0$
$(x - 2)(x + 2) = 0$
$x = 2$ or $x = -2$

$$a^2 - b^2 = (a + b)(a - b)$$

Here is another example.

$x^2 - 25 = 0$
$(x - 5)(x + 5) = 0$
$x = 5$ or $x = -5$

Questions

Solve the following quadratic equations.

a) $x^2 - 1 = 0$ (**Ans:** x = -1 and x = 1)
b) $x^2 - 121 = 0$ (**Ans:** x = -11 and x = 11)
c) $9x^2 - 4 = 0$ (**Ans:** x = -2/3 and x = 2/3)

Solving quadratic equations (regular method)

For example, consider the equation $x^2 + 3x + 2 = 0$.

$$x^2 + 3x + 2 \qquad = 0$$
$$x^2 + x + 2x + 2 \qquad = 0$$
$$x(x + 1) + 2(x + 1) \quad = 0$$
$$(x + 2)(x + 1) \qquad = 0$$
$$x = -2 \text{ and } x = -1$$

Questions

Solve the following quadratic equations.

1. $x^2 - 6x + 8 = 0$ (**Ans:** x = 2 and x = 4)
2. $x^2 + x - 20 = 0$ (**Ans:** x = -5 and x = 4)
3. $x^2 - 4x - 21 = 0$ (**Ans:** x = 7 and x = -3)

Solving linear simultaneous equations (elimination method)

For example, consider the equations $2x + y = 8$ and $x + y = 5$. Solve these simultaneous equations.

Take one equation from the other
$$(2x - x) + (y - y) = (8 - 5)$$
$$x + 0 = 3$$
$$x = 3.$$

Now using $x + y = 5$, we have
$$y = (5 - x) = 5 - 3 = 2$$

Questions

Solve the following simultaneous equations.

1. $x + 3y = 10$ (**Ans:** $x = 1$ and $y = 3$)
    $x + y = 4$
2. $4x - 2y = -4$ (**Ans:** $x = -2$ and $y = -2$)
    $x - y = 0$
3. $2x - 5y = -5$ (**Ans:** $x = 0$ and $y = 1$)
    $x + y = 1$

Solving simultaneous equations (substitution method)

For example, consider simultaneous equations
$y - x = 4$ (1)
$y^2 - x^2 = 8$ (2)

From equation (1)
$y = 4 + x$
Substitute, $y = 4 + x$ into equation (2)
$y^2 - x^2 = 8$
$(4 + x)^2 - x^2 = 8$
$16 + 8x + x^2 - x^2 = 8$
$16 + 8x = 8$
$8x = -8$
$x = -1$

Substitute $x = -1$ in equation (1)
$y - x = 4$
$y - (-1) = 4$
$y = 3$

Questions

Solve the following simultaneous equations.

1. $x + y = 6$ (**Ans:** $x = 2$ and $y = 4$)
    $x^2 - y^2 = -12$

2. $2x + 3y = -8$ (**Ans:** $x = -1$ and $y = -2$)
    $x^2 + y^2 = 3$

Solving simultaneous equations (word problems)

For example, "The sum of two numbers is 12 and their difference is 4. What are those numbers?"

First express the information given in the form of equations. Let the numbers be x and y. Then x + y = 12 and x - y = 4

Using the substitution method
let x + y = 12   (1)
let x - y = 4   (2)

x - y = 4
x = 4 + y

Substitute x = 4 + y into equation (1)
x + y = 12
(4 + y) + y = 12
4 + 2y = 12
2y = 8
y = 4

Substitute y = 4 into equation 1, and therefore x = 8.
The numbers are y = 4 and x = 8.

Question

A store sells books and pens in the following packages. Two books and three pens cost $11 and four books and one pen cost $17. Find the cost of a book and a pen.

(**Ans:** One book cost $4 and one pen cost $1)

Prove two algebraic expressions are identical

Two equations are identical if for any value of x, they have the same value.

Example

Verify the expression $x^2 + 7x + 12 = (x + 4)(x + 3)$.

$x^2 + 7x + 12 = (x + 4)(x + 3)$

when x = 1

$x^2 + 7x + 12 = (1)2 + 7(1) + 12 = 20$
and

$(x + 4)(x + 3) = (1 + 4)(1 + 3) = (5) \times (4) = 20$
Therefore the two equations are identical.

Represent direct and indirect variation symbolically

Let 'k' equal to a constant, if

* y varies directly as x, then y = kx
* y varies directly as the cube of x, then $y = kx^3$
* y is indirectly proportional to x, then y = k/x
* y is indirectly proportional to the square of x, then $y = k/x^2$

Questions

1.   Suppose y is directly proportional to x and when x = 5, then y = 15. What is y when x = 20?

y = kx
15 = k x 5
15 / 5 = k
3 = k

thus for x = 20
y = 3x
y = 3 x 20
y = 60

2.   If y varies inversely proportional to x and y = 4 when x = 12. Calculate

(a) the constant of proportion , k

y is proportional to 1/x, therefore y = k/x

Since y = 4 and x = 12
4 = k/12

k = 4 x 12
k = 48

(b) x when y = 6

Using the answer found in (a), k = 48

y = 48/x
6 = 48/x
6x = 48
x = 48/6
x = 8

Questions

Solve the following problems.

1. If y is inversely proportional to x, and y = 5 when x = 2, find

(i) the constant of proportionality k   (**Ans:** k = 10)
(ii) x when y = 5   (**Ans:** x = 2)

2. If y is proportional to the square of x, and x = 9 when y = 36, what is x when y = 24?   (**Ans:** x = 54)

Questions
Algebra : I
Corey Worrell
Lester Vaughan School
Barbados

> ▸ Video solutions to these questions via the App "**CTS Maths**" for your smartphone or tablet. Details on http://CaribbeanTeachersSeries.com

[**1**] If $x = 2$, $y = -3$ and $z = 4$, calculate $x^2 (3y + z)$

[**2**] Factorize $8x^2 y + 2xy^3$

[**3**] Factorize $20ax - 16ay + 15bx - 12by$

[**4**] Expand and simplify $(2x - 4)(3x + 2)$

[**5**] Simplify $((x + 3) \div 4 + ((x - 6) \div 5)$

[**6**] Simplify $((x^2 y) \div z(((z^3 x) \div (y^2)))$

[**7**] Make R the subject of the equation $T = \sqrt{(x - y)^2 + R^2}$

[**8**] Solve the inequality $4a + 2 \leq 6a - 4$

[**9**] Factorize $4a^2 - 36$

[**10**] Factorize $((a^2) \div (49) - ((b^2) \div ((36))$

[**11**] Factorize $r^2 - t^2 - 6r + 6t$

[**12**] Solve the simultaneous equations $5x + y = 25$ and $2x + y = 13$

[**13**] (i) Solve $x - y = 2$ and $xy = 24$ (ii) Factorize $y^2 + 2y - 24 = 0$

[**14**] Factorize $3x^2 - 10x - 8 = 0$

[**15**] Simplify $\left(\dfrac{2a^7}{3b^4} \right)^3$

[**16**] Simplify $\left(\dfrac{x+y}{x^2 - y^2} \right)$

[**17**] If $x = 2$, $y = 4$, $z = -5$, solve $\dfrac{x^3 + yz}{y^2 - x^2}$

[**18**] The sides of a rectangle are 10 cm and x cm. Find the value of x if the perimeter is 28 cm.

[**19**] Factorize and solve the equation $x^2 + 5x + 6 = 0$

[**20**] Simplify $2\left(\dfrac{3x+2}{3} \right) + 2\left(\dfrac{x-3}{2} \right)$

Chapter 4 : II
Algebra : II
A'ja Maxwell
Queens College
Barbados

Algebra is the study of the unknown. What does that mean? We are often seeking to find some unknown quantity and instead of having to state that quantity each time, we can use a letter or symbol to represent it. Algebra also allows us to look at concepts in a broad sense, make generalisations about things and see patterns of behaviour. For example, we can study the properties of numbers without having to do the same thing with hundreds of numbers. The skills gained from studying Algebra are, therefore, essential and can be applied in almost ALL other areas of academia and life. Make a real effort to master them!

Simplification of algebraic fractions

Simplification of algebraic fractions follows the same rules as the simplification of numeric fractions for the most part.

Addition and subtraction
- Fractions can only be added or subtracted if they have the same denominator.
- If they don't, the LCM of the two denominators must be found.
- Equivalent fractions must then be made out of the LCM.
- Then you can add or subtract the numerators if the terms are alike.

Multiplication and division

- For division problems, invert the second fraction and multiply the first by it.
- Look for opportunities to cancel terms in the numerator with terms in the denominator.
- Multiply across numerators and denominators.
- Reduce the final answer only if ALL of the terms have a common factor.

Examples

Simplify the following.

1. $$\frac{5}{2w}+\frac{8}{9w}=\frac{9\times 5+2\times 8}{18w}=\frac{45+16}{18w}=\frac{61}{18w}$$

2. $$\frac{10t-3r}{4}-\frac{6t+7r}{10}=\frac{5(10t-3r)-2(6t+7r)}{20}=\frac{50t-15r-12t-14r}{20}=\frac{38t-29r}{20}$$

3. $$\frac{7su}{17vx}\times\frac{18vy}{14uz}=\frac{s}{17x}\times\frac{9y}{z}=\frac{9sy}{17xz}$$

4. $$-\frac{a}{2p^7}\div\frac{10a}{27p^2}=-\frac{a}{2p^7}\times\frac{27p^2}{10a}=-\frac{1}{2p^5}\times\frac{27}{10}=-\frac{27}{20p^5}$$

The laws of indices

- $b^x \times b^y = b^{x+y}$
- $b^x \div b^y = b^{x-y}$
- $b^0 = 1$
- $\left(b^x\right)^y = b^{xy}$
- $b^{-x} = \dfrac{1}{b^x}$

> The laws of indices are applied only when the bases are the same and it is only the indices that are changed.

Examples

Simplify the following, leaving no negative indices.

1. $$\frac{c^9 \times c^{11}}{c^{14}}=\frac{c^{9+11}}{c^{14}}=\frac{c^{20}}{c^{14}}=c^{20-14}=c^6$$

2. $\dfrac{2 \times 9^8 \times 2^8}{9^3 \times 2^{12} \times 9^5} = 2^{1+8-12} \times 9^{8-(3+5)} = 2^{-3} \times 9^0 = \dfrac{1}{2^3} \times 1 = \dfrac{1}{8}$

3. $\left(d^{10}e^3\right)^7 = d^{10 \times 7}e^{3 \times 7} = d^{70}e^{21}$

Solving linear equations in 1 unknown

- If the variable appears more than once in the equation, group like terms first.
- If it appears only once, consider what has been done to the variable, and undo each operation in reverse order.

Examples

Solve the following equations.

1. $\dfrac{5f}{8} + 3 = 5$

$\dfrac{5f}{8} = 5 - 3 = 2$

$5f = 2 \times 8 = 16$

$f = \dfrac{16}{5} = 3\dfrac{1}{5}$

2. $87 - g = 3g + 7$

$87 - 7 = 3g + g$

$80 = 4g$

$\therefore g = \dfrac{80}{4} = 20$

> An equation is not solved until you know the value of the variable. (e.g. h = 23)

Solving simultaneous linear equations algebraically

There are two methods – elimination and substitution. Use substitution when at least 1 of the variables has a coefficient of 1 or -1, and use elimination otherwise.

- **Substitution** - Make the variable with a coefficient of 1 or -1 the subject of that equation and then substitute that expression into the other equation and solve.

- **Elimination** – Find the LCM of the coefficients of one of the variables and multiply each equation in order to make the coefficients equal to the LCM. If the coefficients now have the same signs, subtract the two equations to eliminate the variable but if they have different signs, add the two equations instead.

Examples

Solve the following pairs of simultaneous equations:

1. $4i - j = 0$ (1)
      $20i + 3j = -96$ (2)

From (1)

$j = 4i$ (3)

> Make sure you have found the value of BOTH variables before you go on to the next question.

Substitute for j into (2)

$20i + 3(4i) = -96$

$20i + 12i = -96$

$32i = -96$

$\therefore i = \dfrac{-96}{32} = -3$

Substitute for i into (3)

$j = 4 \times -3 = -12$

Ans: $i = -3, j = -12$

2. $7k + 20l = 36$ (1)
      $10k + 7l = 73$ (2)

Multiply (1) x 10
$70k + 200l = 360$ (3)

Multiply (2) x 7
$70k + 49l = 511$ (4)

Thus (3) - (4)
$151l = -151$

$\therefore l = \dfrac{-151}{151} = -1$

Let $l = -1$ in (1)

$7k + 20(-1) = 36$

$7k - 20 = 36$

$7k = 36 + 20 = 56$

$\therefore k = \dfrac{56}{7} = 8$

Ans: $k = 8, l = -1$

3. $7m - 4n = -25$ (1)

       $9m + 10n = 89$ (2)

Multiply (1) x 5

$35m - 20n = -125$ (3)

Multiply (2) x 2

$18m + 20n = 178$ (4)

Thus (3) + (4)

$53m = 53$

$\therefore m = \dfrac{53}{53} = 1$

Let $m = 1$ in (1)

$7(1) - 4n = -25$

$7 - 4n = -25$

$7 + 25 = 4n$

$32 = 4n$

$\therefore n = \dfrac{32}{4} = 8$

Ans: $m = 1, n = 8$

Solving linear inequalities

Solving linear inequalities is exactly like solving linear equations **except** that when the inequality is multiplied or divided by a negative number, the sign changes (from a less than sign (<) to a greater than sign (>) or vice versa).

Examples

Solve the following inequalities and give the solution as a set.

1. $2\left(\dfrac{p-5}{3}\right) \leq 8$ where p is an integer.

$\dfrac{p-5}{3} \leq \dfrac{8}{2}$

$\dfrac{p-5}{3} \leq 4$

$p-5 \leq 4 \times 3$

$p-5 \leq 12$

$p \leq 12+5$

$p \leq 17$

Ans: $\{17, 16, 15, \ldots\}$

2. $-7q+4 > 5-2q$ where q is an integer.

$-7q+2q > 5-4$

$-5q > 1$

$q < -\dfrac{1}{5}$

Ans: $\{-1, -2, -3, \ldots\}$

Changing the subject of formulae

Changing the subject of a formula is similar to solving an equation, treating the variable that you want to make the subject like the variable that you are trying to solve for. In other words, you figure out what operations have been applied to the variable of interest and, unravel them in reverse order. The only difference is that you will be working with other variables now instead of strictly numbers.

Examples

Change the subject of the following formulae to the variable indicated.

1.   Transpose $I = \dfrac{PRT}{100}$ for R.

$100 \times I = PRT$

$100I = R \times PT$

$\dfrac{100I}{PT} = R$

$\therefore R = \dfrac{100I}{PT}$

2.   Transpose $m = \dfrac{y_2 - y_1}{x_2 - x_1}$ to make x_2 the subject.

$m(x_2 - x_1) = y_2 - y_1$

$mx_2 - mx_1 = y_2 - y_1$

$mx_2 = y_2 - y_1 + mx_1$

$x_2 = \dfrac{y_2 - y_1 + mx_1}{m}$

Factorizing as a difference of two squares

First make it clear what the two perfect squares are, and then it is quite easy to put the correct factors into the brackets.

$a^2 - b^2 = (a+b)(a-b)$

Examples

Factorize the following.

1. $4m^2 - 25$

$= (2m)^2 - 5^2$

$= (2m - 5)(2m + 5)$

> Sometimes a common factor must be taken out first in order to see the difference of two squares clearly.

2. $32 - 8a^4$

$= 8(4 - a^4)$

$= 8\left(2^2 - (a^2)^2\right)$

$= 8(2 + a^2)(2 - a^2)$

Factorizing by grouping

Factorizing by grouping problems generally have four terms at the start e.g. $ax + bx + ay + by$. Divide the terms into two pairs which you factorize separately, and then a common factor must emerge out of this step which is placed in one bracket, and the other terms are placed in the other.

Examples

Factorize the following.

1. $px - sx + pm - ms$
$= x(p - s) + m(p - s)$
$= (p - s)(x + m)$

> Sometimes the terms have to be rearranged in order to see the common factor.

2. $-5rt + 2az + 2ar - 5tz$

Both pairs of terms here only have 1 as a common factor, so we can rearrange the terms before factorizing.

$= 2ar + 2az - 5rt - 5tz$
$= 2a(r + z) - 5t(r + z)$
$= (r + z)(2a - 5t)$

> Care must be taken to pay attention to the signs.

Factorizing quadratic expressions

In order to factorize a quadratic of the form $ax^2 + bx + c$:

- Find two numbers whose product is ac and whose sum is b by examining factors of ac in pairs.
- If $a = 1$ put the two numbers directly into the brackets.
- If $a \neq 1$ replace the bx term using the two numbers and factorise by grouping.

Examples

Factorise the following.

1. $p^2 + 13p + 42$
$= (p + 6)(p + 7)$

> Product is 42
> Sum is 13
> Factors in pairs are 1 and 42, 2 and 21, 3 and 14, 6 and 7

2.
$$2m^2 - 9m + 4$$
$$= 2m^2 - m - 8m + 4$$
$$= m(2m-1) - 4(2m-1)$$
$$= (2m-1)(m-4)$$

> Product is 8
> Sum is -9
> Factors in pairs are -1 and -8

3.
$$3n^2 - 7n - 40$$
$$= 3n^2 + 8n - 15n - 40$$
$$= n(3n+8) - 5(3n+8)$$
$$= (3n+8)(n-5)$$

> Product is -120
> Sum is -7
> Factors in pairs are 1 and -120, 2 and -60, 3 and -40, 4 and -30, 5 and – 24, 6 and – 20, 8 and -15

Questions
Algebra : II
Clare Williams
Queens College
Barbados

> ▸ Video solutions to these questions via the App "**CTS Maths**" for your smartphone or tablet. Details on http://CaribbeanTeachersSeries.com

QUESTIONS

[**1**] Solve the following quadratic equations.
(a) $x^2 - 6x + 8 = 0$
(b) $x^2 + x = 12$

[**2**] Solve the quadratic equation $2x^2 + x - 15 = 0$

[**3**] Solve the quadratic equation $x^2 + 7x - 2 = 0$ giving your answer to 2 decimal places.

[**4**] Solve the quadratic equation $3x^2 - 4x = 10$ giving your answer to 2 decimal places.

[**5**] A boy is m years old. His sister is three years younger than the boy. His father is four times older than the sister. If the total of their ages is 57, how old is each person?

[**6**] Ryan and Shawn share 42 marbles. If Ryan gives Shawn 3 marbles then Shawn will have twice as many marbles as Ryan. How many marbles does each boy have?

[**7**] The length of a rectangle is $(2x+1)$ cm and the width is $(x-2)$ cm. If the perimeter of the rectangle is at least 52 cm, then find the minimum value of

(a) x
(b) the width
(c) the length.

[**8**] Mary needs 80g of butter to make a batch of cookies. She uses twice as much to make a batch of frosting. She has a maximum of 1kg of butter which she may use. If she makes 7 batches of cookies, what is the maximum number of batches of frosting she can make?

[**9**] Chad buys pencils and markers. He buys 11 items in total. The cost of a pencil is $0.70 and the cost of a marker is $1.50. He spends a total of $10.90. Find the number of pencils and the number of markers that he buys.

[**10**] A supermarket has two shuttle vehicles. Vehicle 1 can carry x passengers and vehicle 2 can carry y passengers. On one day, the first vehicle made 8 journeys and the second made 12 journeys carrying a total of 168 passengers. On the next day, the first vehicle made 4 journeys and the second made 1 journey carrying a total of 34 passengers.

(a) Form two equations in x and y
(b) Solve the equations to find the number of passengers that each vehicle can carry.

[**11**] A triangle has a base of $(2x+1)$ cm and a perpendicular height of $(2x-8)$ cm. The area of this triangle is 45cm^2. Form an equation in x and solve it to find the value of x.

[**12**] The sum of the squares of three consecutive integers is 110. Form a suitable quadratic equation and solve it to find the value of the three numbers.

[**13**] Solve the equations $x+y=8$ and $xy+3y=30$

[**14**] Solve the equations $x+3y=7$ and $x^2+y^2=5$

[**15**] Prove the identity $(x+3)^2 - (x-3)^2 = 12x$

[**16**] Prove the identity $x^3 - x = (x-1)x(x+1)$

[**17**] Express these statements symbolically.
(a) the variables p and q are related such that "p varies directly as $(4q-3)$".
(b) the variables m and n are related such that "m varies inversely as $(2n-5)$".

[**18**] The variables u, v and w are related such that "u varies directly as v^2" and "v varies directly as w^3". Determine how u varies with the respect to w.

[**19**] The variables c and d are related such that "d varies directly as (c^2+5)".

(a) Write an equation in c, d and k to describe this variation, where k is the constant of variation.
(b) if $d=9$ when $c=2$, calculate the value of k.
(c) Calculate the value of d when $c=5$
(d) Calculate the value of c when $d=54$

[**20**] The variables d and t are related such that "t varies inversely as $(4d-1)$".

(a) Write an equation in d, t and k to describe this variation, where k is the constant of variation.

(b)

d	3	13	b
t	10	a	4

Using the information in the table above, calculate the value of
(i) k, the constant of variation
(ii) a
(iii) b

Chapter 5

Sets

Patrick Cadogan

Harrison College

Barbados

Welcome learners to the topic, **Set Theory**. The focus will be on problem solving involving two sets and three sets respectively. Firstly, this chapter will define the common *terms and expressions* seen and used in Set Theory.

Secondly, the *techniques for problem solving* will be explored. It is hoped that students will have a solid understanding of the terminologies used and it is expected that students will possess a good grasp of basic algebraic concepts.

Terms commonly encountered in Set Theory

Term	Meaning
Term	**Meaning**
Calculate	To compute (or work out).
Describe	To write down using words, or using a combination of words and mathematical symbols.
Determine	To deduce and or calculate.
Disjoint	Two or more sets that are distinct and separate.
Expression	A single term or collection of terms connected by the mathematical operators addition, subtraction, multiplication or division.
Equation	Two mathematical statements or expressions separated by an equal sign.
List	A collection of terms typically written down in a sequential order and enclosed by brackets.
Set	A collection of members possessing similar properties.
Set Notation	Mathematical symbols for describing a set or description of a relationship between sets.
Set Builder Notation	A convenient way of listing the elements (or members) of an entire set without having to name or identify each element of the set. Note, inequality symbols are a common feature of this notation.
A Finite Set	A set in which each of the elements (or members) can be listed or stated. The number of elements in this set can be counted.
An Infinite Set	A set in which each of the elements (or members) cannot be listed or stated. The number of elements in this set cannot be counted.
Equal Sets	Two or more sets containing the same number of elements (or members), and significantly, the elements of the sets are identical. The order of the elements in each set is unimportant.

Terms and their symbols commonly encountered in set theory

Term	Symbol	Meaning
At least	≥	Greater than or Equal to
At most	≤	Less than or Equal to, OR, not more than
The cardinality of set A	n(A)	The number of elements (or members) in Set A
Complement	′	Members of the Universal Set which are excluded from a particular set or combination of sets
Intersection	∩	Members common to two or more sets
Is a member of	∈	Belonging to a particular set
Is not a member of	∉	Not belonging to a particular set
Is a subset of	⊆	Either of; any of the elements of the set or combination of elements of the set, the empty set (denoted by { } or ϕ), or the set itself
Is not a subset of	⊄	None of the subsets of the given set
Is a proper subset of	⊂	Either of; any of the elements of the set or combination of elements of the set, but excluding the empty set and the set itself
Is not a proper subset of	⊄	None of the proper subsets of the given set
Union	∪	The joining together of the members of two or more sets
Universal Set	**U**	A set containing all possible elements and all associated subsets

Set theory is best learned by example. The following examples will bring out certain features to help clarify key concepts.

U = {1, 2, 3, …, 20} and given that sets A and B are both subsets of U such that A = {multiples of 2} and B = {multiples of 3}. List the elements of the following sets:

(i) A' (ii) B' (iii) $A \cap B$ (iv) $A \cup B$ (v) $(A \cap B)'$
(vi) $(A \cup B)'$ (vii) $A' \cap B'$ (viii) $A' \cup B'$ (ix) $A' \cap B$ (x) $A' \cup B$

Solution

Firstly, we can list the elements of both sets A and B as the number of elements of each set is low.
A = {2, 4, 6, 8, 10, 12, 14, 16, 18, 20} and B = {3, 6, 9, 12, 15, 18}.

(i) A' = {1, 3, 5, 7, 9, 11, 13, 15, 17, 19}
(ii) B' = {1, 2, 4, 5, 7, 8, 10, 11, 13, 14, 16, 17, 19, 20}
(iii) $A \cap B$ = {6, 12, 18}
(iv) $A \cup B$ = {2, 3, 4, 6, 8, 9, 10, 12, 14, 15, 16, 18, 20}
(v) $(A \cap B)'$ = {1, 2, 3, 4, 5, 7, 8, 9, 10, 11, 13, 14, 15, 16, 17, 19, 20}
(vi) $(A \cup B)'$ = {1, 5, 7, 11, 13, 17, 19}
(vii) $A' \cap B'$ = {1, 5, 7, 11, 13, 17, 19}
(viii) $A' \cup B'$ = {1, 2, 3, 4, 5, 7, 8, 9, 10, 11, 13, 14, 15, 16, 17, 19, 20}
(ix) $A' \cap B$ = {3, 9, 15}
(x) $A' \cup B$ = {1, 3, 5, 6, 7, 9, 11, 12, 13, 15, 17, 18, 19}

Note the following facts:

- $(A \cap B)' = A' \cup B'$
- $(A \cup B)' = A' \cap B'$

In a survey of a group of students it was recorded that 105 play Sports (S), 59 play Video Games (V), x play both and 18 play neither.

a)   Draw a Venn diagram to illustrate this information.
b)   Write an expression in terms of x for the **total** number of students in the survey.
c)   Given that 160 students were surveyed, form an equation and use it to calculate the number of students who play both Sports and Video Games.

Solution

a)   Firstly, we fill in the intersection when completing the Venn diagram. Secondly, we subtract the amount in the intersection from the total of each of the two given sets. Note, the number of students in the complement of S union V is 18, that is, $n(S \cup V)' = 18$.

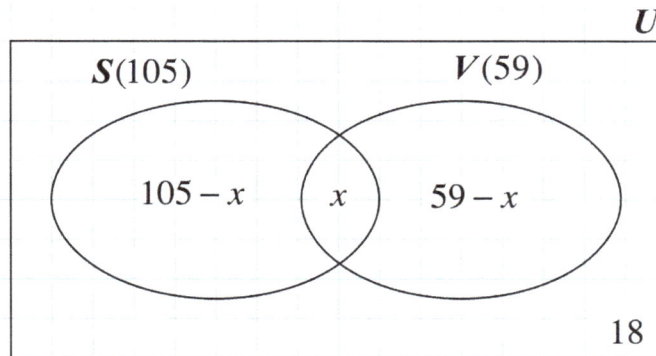

b)   Required expression is $(105 - x) + x + (59 - x) + 18$, this simplifies to $182 - x$.

c)   Equation to be solved is $182 - x = 160 \rightarrow 182 - 160 = x \rightarrow 22 = x$. Therefore, 22 students play both Sports and Video Games.

Note
- number who play Sports only is $(105 - x) = 105 - 22 = 83$
- number who play Video Games only is $(59 - x) = 59 - 22 = 37$

Example 3 : Set Notation

A set P from the set of natural numbers, N, is such that each unique member is at least greater than 5 but not more than 15.

> Learners must be familiar with the sets of numbers commonly encountered in set theory, primarily the natural numbers N (or counting numbers), the whole numbers W, and the integers Z.

a)   Using set notation, write down a relationship between the two sets and an algebraic expression to represent set P.

(b)   Determine the number of members in set P.

Solution

a)   We can deduce that **P** is a proper subset of **N**, so we write **P ⊂ N**.

We will use *x* to represent each member of the set **P**. So in words, we can say that **P** is the set of all *x* such that *x* is a member of the set of natural numbers, *x* being greater than or equal to 5 but less than or equal to 15. Using appropriate mathematical symbols, we write **P** is $\{x: x \in N; 5 \leq x \leq 15\}$

b)   Counting from 5 to 15 inclusive, we write $n(P) = 15 - 5 + 1 = 11$

Example 4 : Unknown Variable

In a group of persons it was found that *x* eat vegetables (**V**), 30 eat chicken (**C**) and 28 eat fish (**F**). Secondly, 12 eat both vegetables and chicken, 17 eat both vegetables and fish, and 9 eat both chicken and fish. Thirdly, 5 persons eat all three food items and each person eats at least one of these foods.

a)   Copy and complete the Venn diagram shown to illustrate **all** of the above information.

b)   Write down an expression in *x* for the number of persons who eat vegetables **only** or chicken **only** but not both.

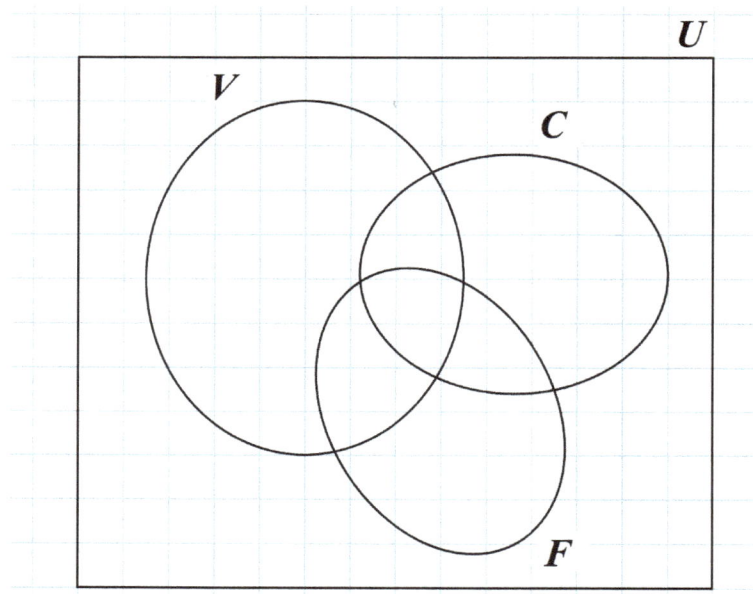

Solution

a)   We will begin by filling in the intersection of all three sets, followed by the intersections involving two sets. The number of persons who eat one type of food **only**, will be found by subtracting the total of the triple and double intersections, from the overall total of each particular food item. Since each person eats at least one of the foods, this implies that the complement of **V** union **C** union **F** is 0, that is, $n\,(V \cup C \cup F)' = 0$.

Now 5 persons eat all three food items, so

- **V** and **C** only is 12 – 5 = 7
- **V** and **F** only is 17 – 5 = 12
- **C** and **F** only is 9 – 5 = 4
- **V** only is $x - (7 + 5 + 12) = x - 24$
- **C** only is 30 – (7 + 5 + 4) = 14
- **F** only is 28 – (12 + 5 + 4) = 7

Inserting this information in the Venn diagram, we have

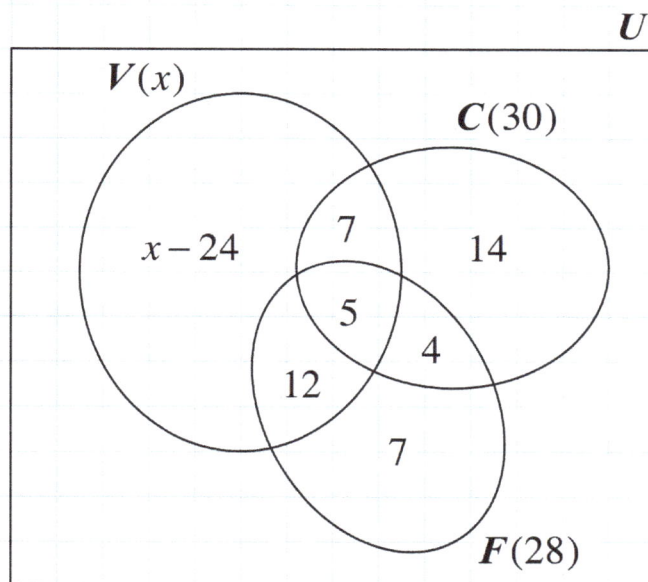

(ii) From the Venn diagram, the number of persons who eat vegetables **only** is given by $x - 24$, and the number who eat chicken **only** is 14. Therefore, the number of persons who eat vegetables **only** or chicken **only** but not both is obtained from the sum of $x - 24$ and 14 which equals $x - 10$.

Questions
Sets
Patrick Cadogan
Harrison College

Barbados

> ▸ Video solutions to these questions via the App **"CTS Maths"** for your smartphone or tablet. Details on http://CaribbeanTeachersSeries.com

QUESTIONS

[1] The Venn diagram below shows two sets P and Q.

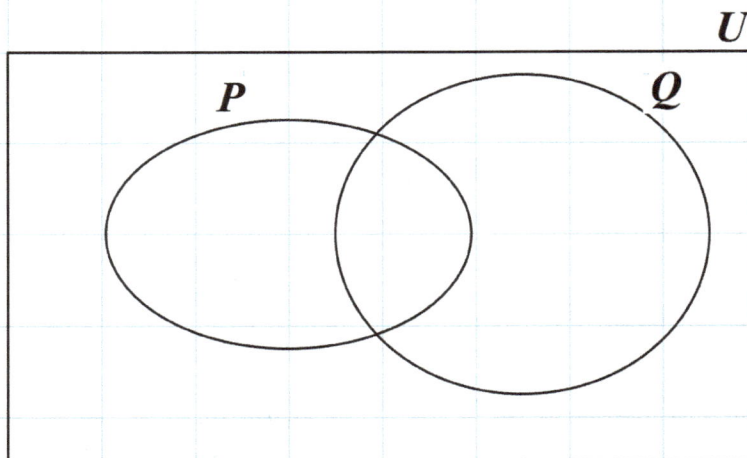

$n(P) = 35$, $n(Q) = 42$, $n(P \cap Q) = x$ and $n(P \cup Q)' = 0$.

i)   Copy and complete the Venn diagram to show this information.
ii) If $n(P \cup Q) = 64$, calculate the value of x.
iii) Hence, find $n(P' \cap Q)$.

[2] The information below was obtained from a group of students.

- 14 own a bicycle (B) only
- x own a scooter (S) only
- 12 own both a bicycle and a scooter
- $3x$ own neither and this amount is equal to the number who own a scooter

Represent this information in a Venn diagram. Hence, calculate (a) the value of x and (b) the number of students in the group.

[3] In the diagram below, the universal set U represents all the athletes in an athletics club. The set R represents the 36 athletes whose specialty is running, while the set J represents the athletes whose specialty is jumping. Calculate the

a)   number of athletes whose specialty is jumping.
b)   total number of athletes in the club.

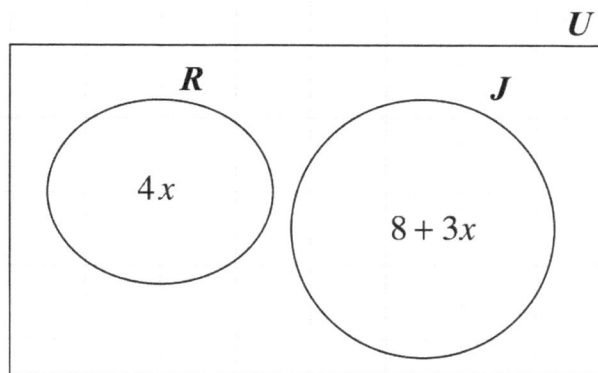

[4] The universal set, U = {Natural numbers less than 12}. L and M are subsets of U, with L = {Multiples of 2 from 2 to 10 inclusive} and set M = {Factors of 8}.

i)   Draw a Venn diagram to show this information.
ii) Hence, list the members of the set,
   a) $(L \cap M)'$
   b) $L \cap M'$
   c) $(L \cup M)'$

[5] The following information was obtained from a survey of 55 students.

- 27 eat broccoli (B)
- 25 eat carrot (C)
- x eat both broccoli and carrot
- $2x$ eat neither

i)   Draw a Venn diagram to illustrate this information.
ii) Write an expression in terms of x for the total number of students in the survey.
iii) Hence, calculate the number of students who eat (a) both broccoli and carrot (b) neither broccoli nor carrot.

[6] The foreign language club has 60 members who each speak either French (F) or Spanish (S), or both. Specifically, 35 speak French, 32 speak Spanish and x speak both languages.

i)   Draw a Venn diagram to represent this information.
ii) Calculate the number of bilingual members (that is, students who speak both languages).
iii) On your Venn diagram, shade the region which represents $F \cap S'$

[7] On a tour by a group of students to the juice company

- 65 sampled apple juice (A)
- 70 sampled mango juice (M)
- x sampled both types of juice
- $3x$ sampled neither

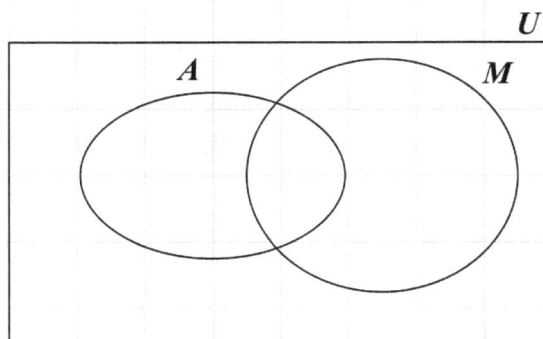

i)   Copy and complete the Venn diagram to illustrate this information.
ii) Given that 15 students sample neither apple juice nor mango juice, determine the number of students who sampled (a) apple juice only (b) mango juice only.
iii) Hence, calculate the total number of students on the tour.

[8] 140 students wrote examinations in Mathematics (M) and English (E). 100 passed Mathematics, 98 passed English, x passed both subjects and 5 passed neither.

i)   Draw a Venn diagram to show the number of students in each subset.
ii) Hence, determine a) $n(M \cap E)$ b) $n(M' \cap E)$

[9] In the Venn diagram below $U = \{x : 1 < x < 12, x \in w\}$. P and Q are subsets of U.

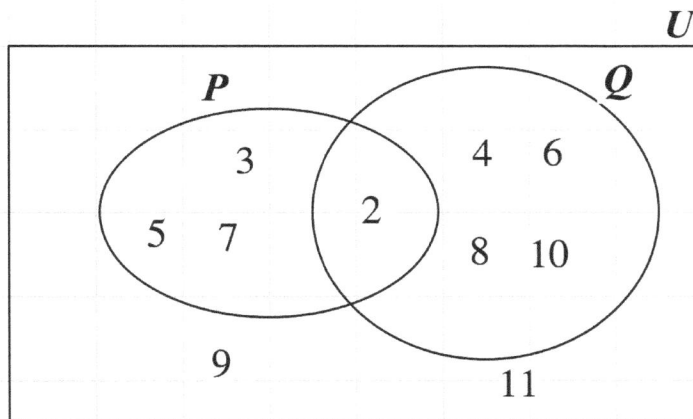

i)   Describe concisely, in words P and Q
ii) Determine $n(P' \cap Q')$

[10] In a group of 46 students

- 28 study History (H)
- $3x$ study Geography (G) only
- x study both History and Geography
- 6 study neither

i)   Draw a Venn diagram to show this information.
ii) Hence, calculate the number of students (a) who study Geography (b) who study only one of the two subjects.

[**11**] Describe the shaded region in the Venn diagram below,

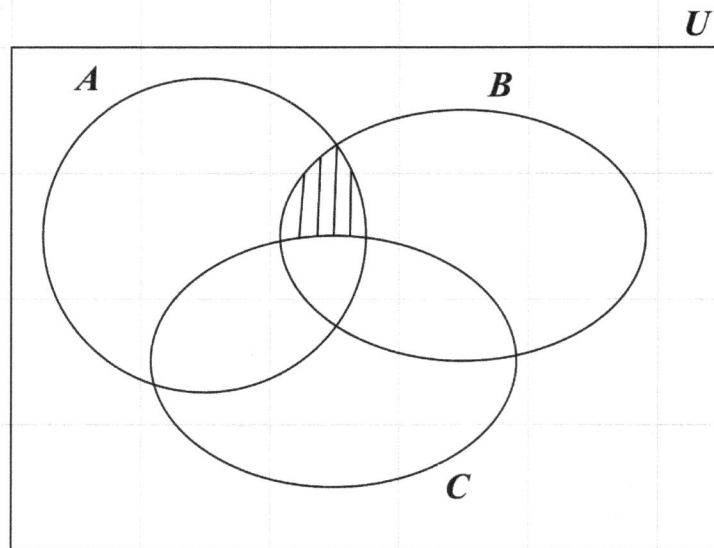

(i) using clear and concise words.
(ii) using set notation.

[**12**] Describe the shaded region in the Venn diagram below,

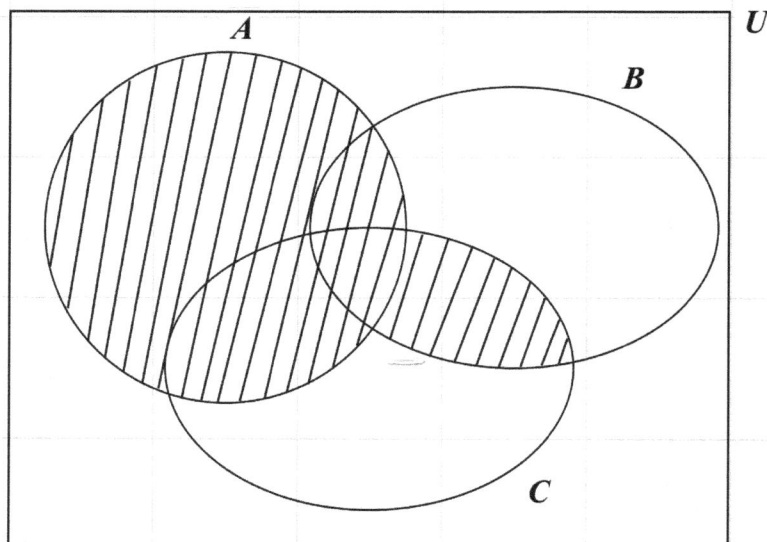

(i) using clear and concise words.
(ii) using set notation.

[**13**] Describe the shaded region in the Venn diagram below,

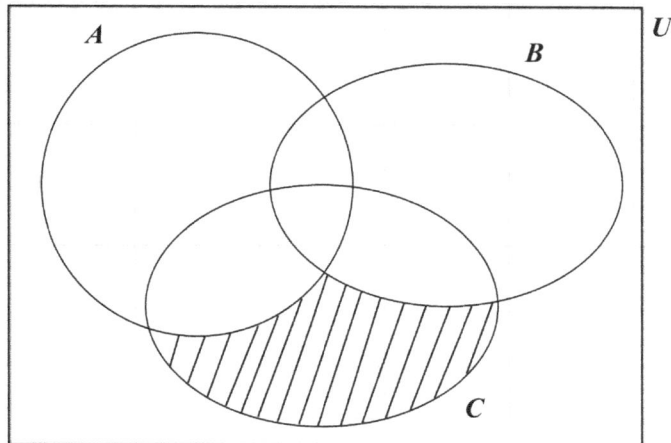

(i)   using clear and concise words.
(ii) using set notation.

[**14**] The universal set $U=\{p,q,r,s,t,u,v,w,x,y,z\}$ and set $A=\{p,q,r,u,v,x\}$ $B=\{p,r,v,w,y\}$ and $C=\{q,u,v,w,z\}$

(i)   Draw a Venn diagram showing the elements in the sets U, A, B and C
(ii) List the numbers of the sets $A\cap(B\cup C)$ and $A\cap(B\cup C)'$
(iii) Determine $n(A\cap B\cup C)'$

[**15**] 92 students were surveyed and the Venn diagram below shows their favourite sporting activities of Cricket (C), Athletics (A) and Netball (N).

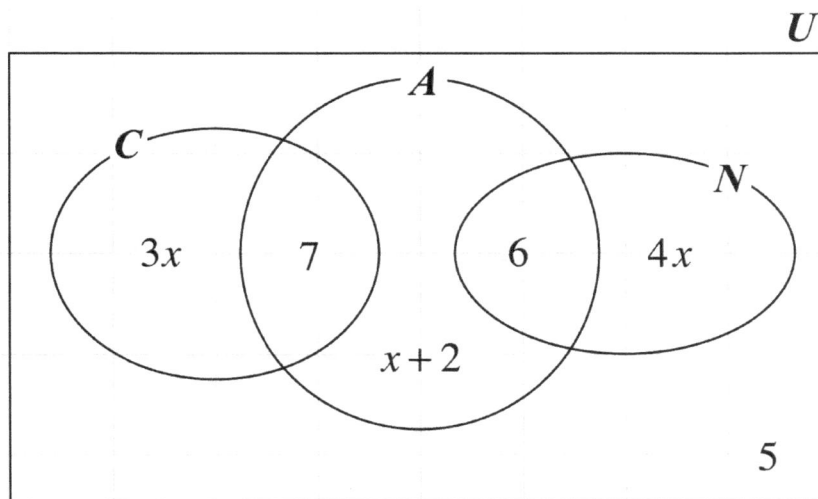

(i)   Write an expression in terms of x to show the number of students in the survey.
(ii) Hence, calculate the number of students whose favourite sport is (a) Cricket (b) Athletics (c) Netball only.

[**16**] The Venn diagram below shows information about the three subjects of Chemistry (C), Biology (B) and Mathematics (M) studied by students at a evening institute. 36 students study Chemistry, and all students study at least one of the three subjects.

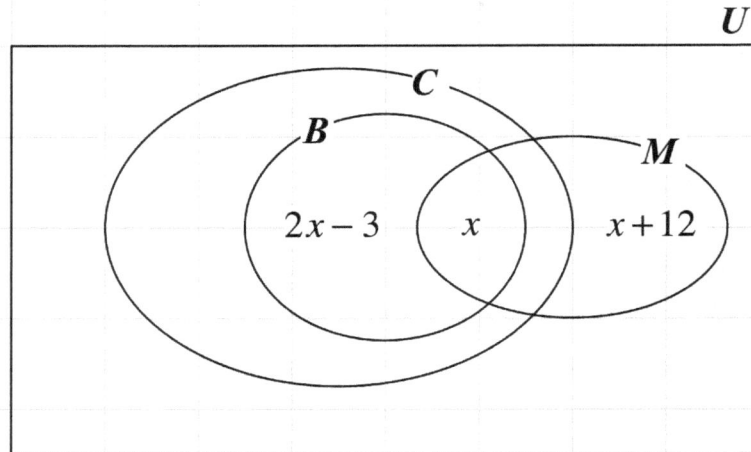

(i)   Describe clearly and concisely in words, (a) the relationship between the members represented by sets B and C and (b) the members of the set $M' \cap C$
(ii) Calculate the number of students who study Mathematics.

[**17**] In a group of students,

- 28 study Mathematics
- 10 study Mathematics only
- 9 study Art only
- 12 study Literature only
- $2x$ study Mathematics and Art only
- $3x$ study Mathematics and Literature only
- 8 study Art and Literature only
- x study all three subjects
- 6 study none of these subjects

(i)   Draw a Venn diagram to illustrate the information.

(ii) Write an equation in x to show the number of students who study Mathematics.

(iii) Hence, calculate the number of students who study (a) all three subjects and (b) the number of students in the group.

[18] 120 students were surveyed and the results are as follows:

- 60 play both the piano (P), 75 play the guitar (G) and 52 play the violin (V)
- 42 play both the piano and the guitar
- 15 play both the piano and violin
- 25 play both the guitar and the violin
- x can play all three instruments and 5 play none of these instruments

(i)   Draw a Venn Diagram to illustrate this information.

(ii) Write and expression in terms of x to show the number of students surveyed.

(iii) Hence, calculate the number of students who can play (a) all three instruments and (b) one instrument only.

[19] In a class of 56 students, 28 wear spectacles (S), 32 wear a watch (W) and 20 wear braces (B), 12 wear spectacles and a watch, 10 wear spectacles and braces, and 9 wear a watch and braces, x students wear all three items and each student wears at least one of these items.

(i)   Copy and complete the Venn diagram below to illustrate this information.

(ii) Write an expression in terms of x to show the number of students in the class.

(iii) Hence, calculate the number of students who wear all three items.

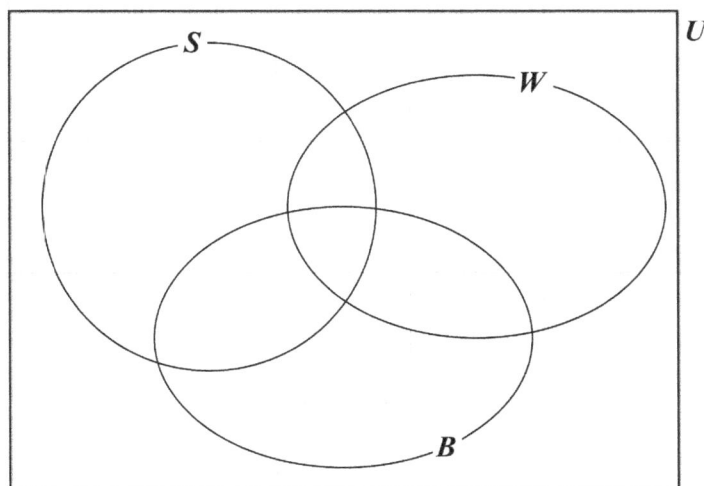

[20] The composition of the membership of a sports club is shown in the Venn diagram below.

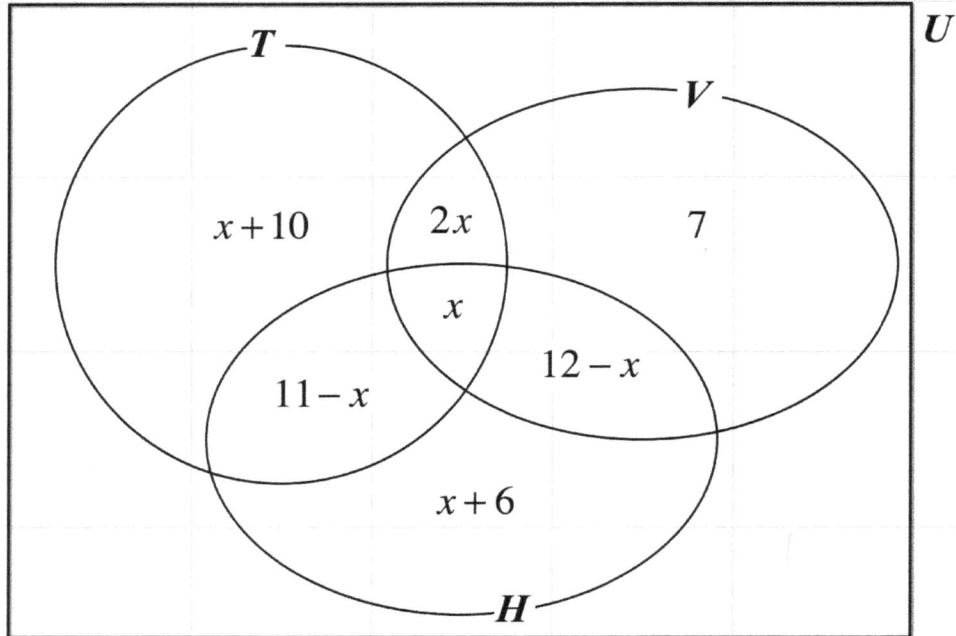

- 25 members play volleyball (V)
- Each member plays at least one of the three sports of volleyball, tennis (T) and hockey (H)

(i)   Using set notation, describe the shaded region.
(ii) Calculate the number of members who play (a) all three sports (b) tennis only and (c) evaluate the total number of members in the club.

Chapter 6
Measurement
Renaldo Holder
Alexandra School
Barbados

Measurement is an important concept in solving problems involving length, area and volume. It is a very wide area in the field of mathematics. Therefore, in this chapter we analyse the five sections of measurement which are,

- Length
- Mass
- Time
- Temperature
- Speed

Length

Length is the distance from one end to the other end of an object. For example, the figure below shows a rectangle with its labeled sides.

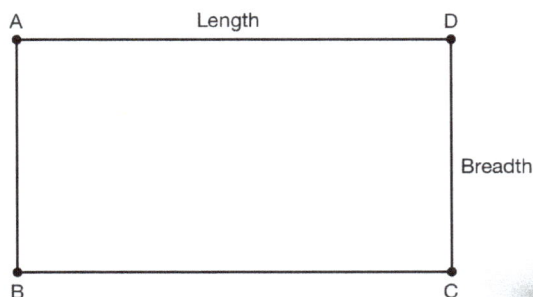

In a rectangle, the length is longer than the breadth. The length is a one dimensional quantity. The standard unit of length is the metre, m

Metric units of length are all related such as

10 mm = 1 cm

100 cm = 1 m

1000 m = 1 km

Examples

Convert the following.

1.   2.1 m to cm

1 m = 100 cm

therefore, 2.1 m = 2.1 x 100 = 210 cm

2.   520 mm to cm.

1 cm = 10 mm

therefore, 520 mm = 520 ÷ 10 = 52 cm

Perimeter

Perimeter is the distance around a plane figure. A plane figure is a geometric shape whose surface is flat. Examples of plane figures are shown below.

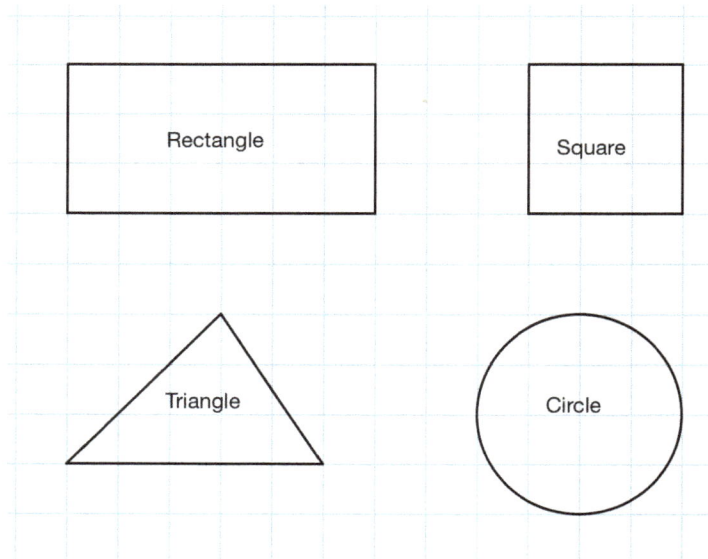

To determine the perimeter of a shape you must know the length of each side of the plane figure. Sometimes, you will have to find a missing length using the perimeter of the shape.

Examples

1. Calculate the perimeter of the rectangle below. The unit is centimetres.

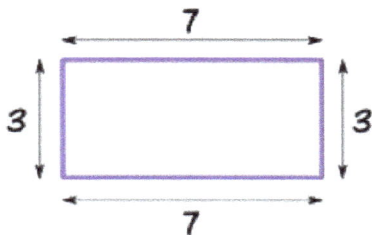

The perimeter is the distance around a plane figure. Thus

$perimeter = 7cm + 7cm + 3cm + 3cm = 20cm$

2. Calculate the perimeter of a rectangle with a length of 5m and a breath of 12m.

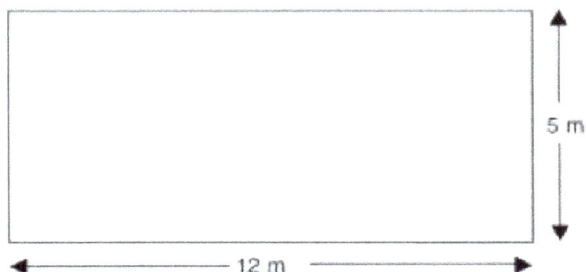

> Draw the shape with its dimensions first to make the question easier to solve. Avoid doing it in your head.

From the properties of a rectangle opposite sides are equal
Hence, $perimeter = 12m + 12m + 5m + 5m = 34m$

Perimeter of a circle

The perimeter of a circle is called the circumference. It is the distance around a circle. The circumference of a circle is about three times its diameter. Hence, this value is called "pi" (.i.e.).

- Circumference - C
- π - is equal to 3.142 or approximately $\frac{22}{7}$
- r - radius
- d - diameter

The equation for the circumference of a circle is
$C = 2\pi r = \pi d$ because $d = 2r$

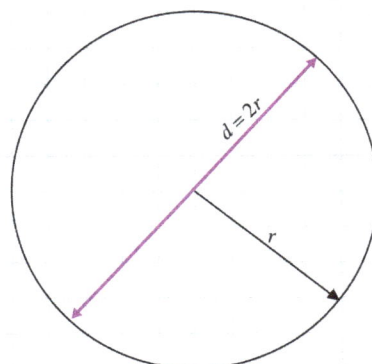

Examples

1.   The radius of a circle is 21 cm. Find the circumference of the circle.

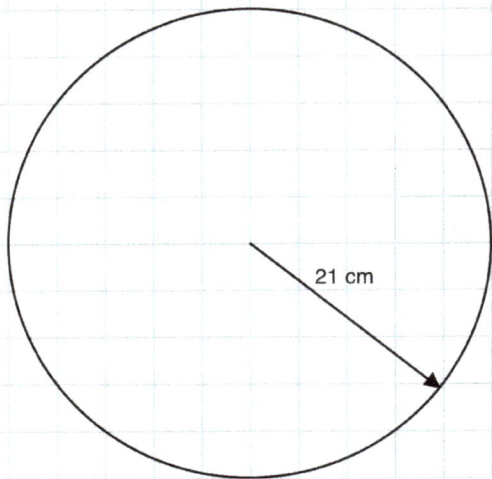

21 cm

Draw the shape with its dimensions first to make the question easier to solve. Avoid doing it in your head.

$$C = 2\pi r = 2 \times \pi \times r = 2 \times \frac{22}{7} \times 21cm = 66cm$$

The circumference is 66 cm. Since the circumference is a distance, our answer must have a measurement quantity assigned to it.

2.   The circumference of a circle is 110 cm. Find the radius of the circle.

The circumference is
$C = 2\pi r$

$110 = 2(3.14)r$

$110 = 6.28r$

$110 \div 6.28 = r$

$17.52 = r$

Since the question asks to find the radius, it is an unknown

Thus, the radius of the circle is 17.52 cm to 2 d.p.

Arc length and sector area

Any part of the circumference of a circle is called an **arc**. The diagram below is a circle showing the arc length of a circle.

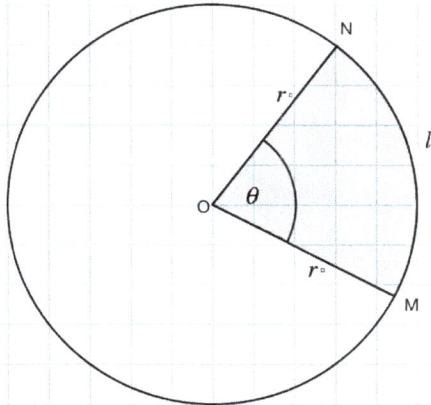

> Arc length is the length from point N to M along the circle

- The arc length l of a circle is given by $l = \dfrac{\theta}{360} \times 2\pi r$

- The area A of a sector is given by $A = \dfrac{\theta}{360} \times \pi r^2$

where θ is the angle subtended by the arc at the centre of the circle.

Examples

1.   An arc subtends an angle of 45° at the centre of a circle of 9cm. Determine the following:

(a)   The length of the arc MN
Arc length l is

$l = \dfrac{\theta}{360} \times 2\pi r$

$= \dfrac{45}{360} \times 2 \times 3.14 \times 9$

$= \dfrac{1}{8} \times 56.52$

$= 7.065\,cm$

$= 7.1\,cm$

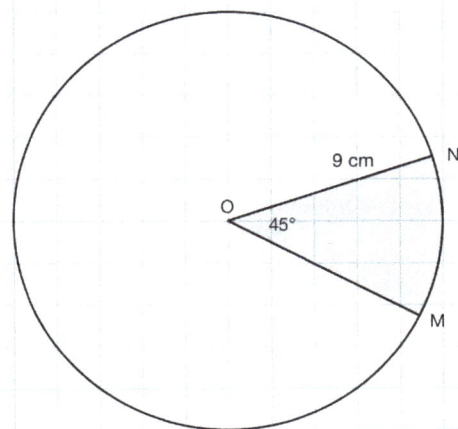

(b)   The area of the sector formed by the arc OMN.

Area of a sector is

$$A = \frac{\theta}{360} \times \pi r^2$$

$$= \frac{45}{360} \times 3.14 \times (9cm)^2$$

$$= \frac{1}{8} \times 3.14 \times 81cm^2$$

$$= \frac{1}{6} \times 254.34cm^2$$

$$= 31.7925cm^2$$

$$= 31.8cm^2$$

2.   The diagram below, not drawn to scale, shows the sector of a circle with centre O with $P\hat{O}Q = 60°$ and $OP = 15$ cm.
Calculate

(a)   The length of the minor arc PQ.

$$l = \frac{\theta}{360} \times 2\pi r$$

$$= \frac{60}{360} \times 2 \times 3.14 \times 15cm$$

$$= \frac{1}{6} \times 94.2cm = 15.7 \ cm$$

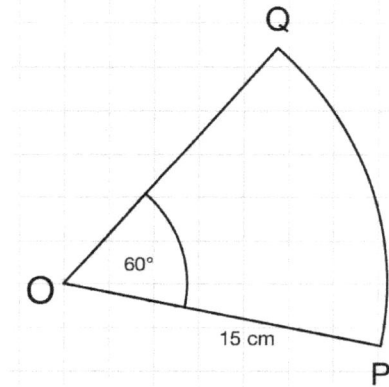

(b)   The perimeter of the POQ.
$$perimeter = PQ + OQ + OP$$
$$= 15.7cm + 15cm + 15cm = 45.7cm$$

(c)   The area of the figure POQ.

$$Area \ of \ POQ = \frac{\theta}{360} \times \pi r^2$$

$$= \frac{60}{360} \times 3.14 \times (15cm)^2$$

$$= \frac{1}{6} \times 706.5cm^2 = 117.775cm^2$$

$$\cong 117.8cm^2$$

Area definition

The **area** is the amount of space inside the boundary of a flat surface. In other words, it can be seen as the number of square units in the dimensional figure.

> The area is found for a 2-dimensional object only which is plane surface.

Area of a square

A square is a plane surface with the length and breadth being equal (see figure below).

- The formula for the area of a square is $A = l \times l = l^2$, where l is the length of one side.

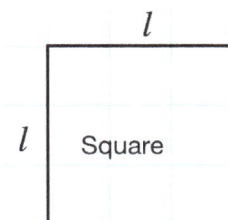

l

l   Square

Examples

1.   Calculate the area of a square which has a length 6cm.

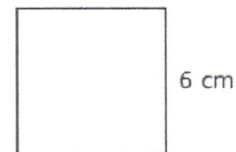

6 cm

Since the length of all the sides of a square are equal,
$A = 6cm \times 6cm = 36cm^2$

2.   If the area of a square is 169 m², find its length l and perimeter in cm.

$A = l \times l = l^2$

$169m^2 = l^2$

$\sqrt{169m^2} = l$

$13m = l$

$perimeter = 13m + 13m + 13m + 13m = 52m$

Area of a rectangle

A rectangle is a plane surface with the length **not** equal to the breadth.

- The formula for the area of a rectangle is $A = l \times b$, where l is the length and b is the breadth.

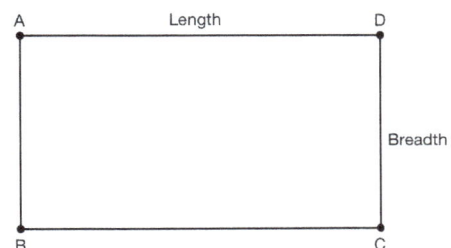

A   Length   D

Breadth

B   C

Examples

1.   A wall has a length of 10 m and a breadth of 6 m.

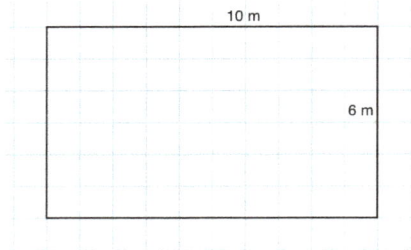

The breadth of the floor is its width.

(a)   Find the area of the wall.

$A = 10m \times 6m$

$= 60m^2$

$m \times m = m^2$

(b)   Find the perimeter of the wall.

$perimeter = 10m + 10m + 6m + 6m = 32m$

2.   The area of a rectangle floor is 144 m² and it length is 16 m. Find its width.

$A = l \times b$

$144m^2 = 16m \times b$

$\dfrac{144m^2}{16m} = b$

$9m = b$

By the law of indices

$m^2 \div m^1 = m^{2-1} = m^1 = m$

3.   A rectangular garden is 16 m in length by 11 m in width. Inside is a path 3 m in the width around the boundaries. Determine the area of the path.

$Area\ of\ outer = 16m \times 11m = 176m^2\quad Area\ of\ inner = 10m \times 5m = 50m^2$

$Area\ of\ path = 176m^2 - 50m^2 = 126m^2$

Parallelogram

A parallelogram is a plane surface which is a quadrilateral whose opposite sides are parallel and equal and opposite angles are equal.

- The formula for the area of a parallelogram is $A = b \times h$, where b is the length of the base and h is the height.

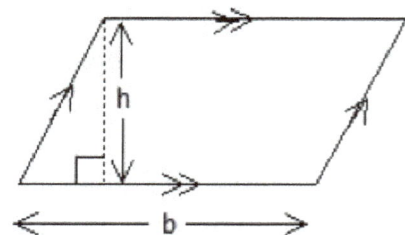

Examples

1.   Determine the area of the parallelogram shown below.

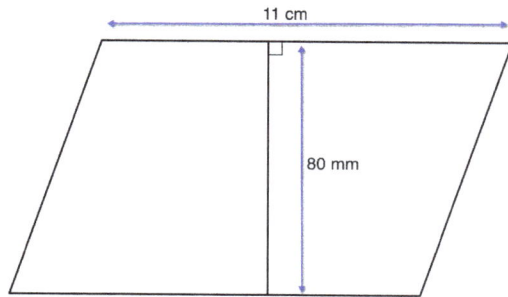

Before we find the area of the parallelogram, we need to have the same units.

Since $1cm = 10mm$, $80mm = 80 \div 10 = 8cm$

$A = b \times h$

$= 11cm \times 8cm = 88cm^2$

Trapezium

A trapezium is a plane surface which is a quadrilateral with only one pair of parallel sides.

• The formula for the area of a trapezium is $A = \left(\dfrac{a+b}{2}\right) \times h$

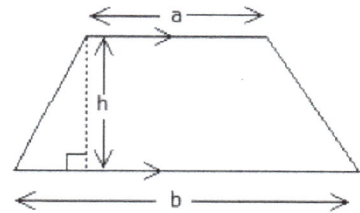

• a is the length of one of the parallel side of the trapezium.
• b is the length of second parallel side of the trapezium.
• h is the perpendicular height (i.e. distance between the parallel lines) of the trapezium.

Examples

1.   Determine the area of trapezium below.

All lengths in cm

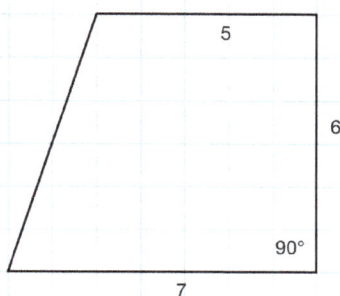

First identify the parallel lengths, that is, 7cm and 5cm and the height which is 6cm.

$$A = \left(\frac{7cm + 5cm}{2} \right) \times 6cm$$

$$= \left(\frac{12cm}{2} \right) \times 6cm$$

$$= 6cm \times 6cm$$

$$= 36cm^2$$

2.   Determine the area of trapezium below.

$$A = \left(\frac{7cm + 4cm}{2} \right) \times 10cm$$

$$= \left(\frac{11cm}{2} \right) \times 10cm$$

$$= 5.5cm \times 10cm$$

$$= 55cm^2$$

All lengths in cm

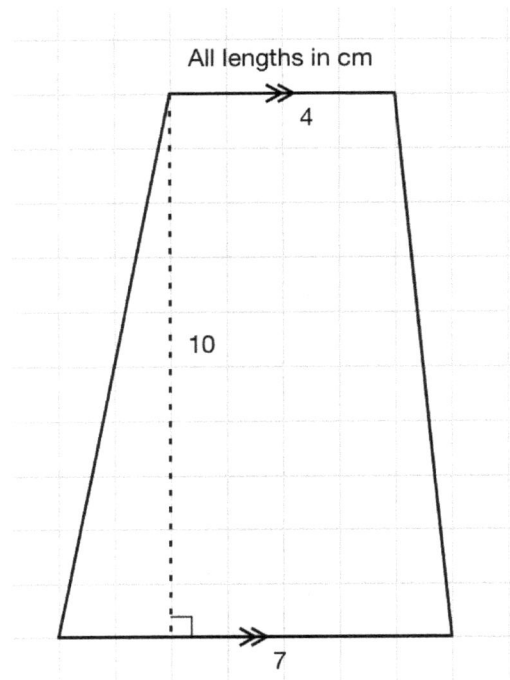

4

10

7

3.   If the area of the trapezium below is 30 cm², what is the value of L cm?

$$30cm^2 = \left(\frac{8cm + L}{2} \right) \times 6cm$$

$$\frac{30cm^2}{6cm} = \left(\frac{8cm + L}{2} \right)$$

$$5cm = \left(\frac{8cm + L}{2} \right)$$

$$5cm \times 2 = 8cm + L$$

$$10cm - 8cm = L$$

$$2cm = L$$

L

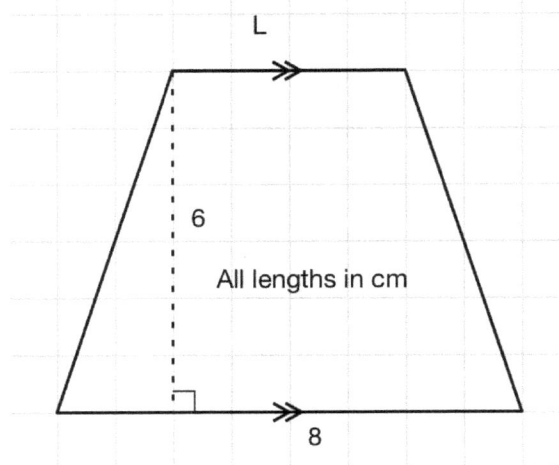

6

All lengths in cm

8

Kite

A kite is a plane surface which is a quadrilateral having two pairs of adjacent sides equal. The diagonals are perpendicular but one of them bisects the kite.

• The formula for the area of a kite is $A = \left(\dfrac{d_1 \times d_2}{2} \right)$

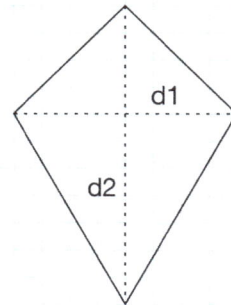

Example

The figure below, not drawn to scale, shows a kite with diagonals AC = 11 cm and BD = 4 cm. Determine the area of the kite.

$d_1 = 4cm$ and $d_2 = 11cm$

$$A = \left(\frac{d_1 \times d_2}{2} \right)$$
$$= \left(\frac{4cm \times 11cm}{2} \right) = \frac{44cm^2}{2} = 22cm^2$$

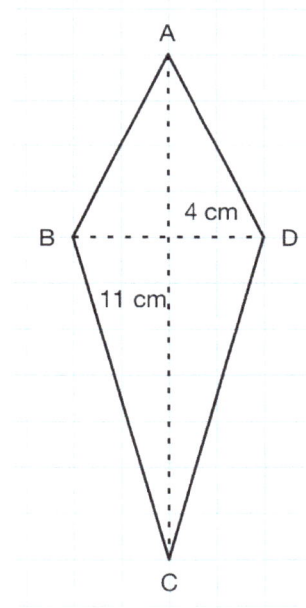

Triangle types

There are six types of triangles.

1.   An equilateral triangle has all sides and angles equal.
2.   An isosceles has two of its sides are equal. Also, the angles opposite the equal sides are equal.
3.   A scalene triangle has no equal sides.
4.   An acute-angled triangle has only acute angles (i.e. less than 90°).

5.   A right-angled triangle has one angle equal to 90°.
6.   An obtuse angled triangle has one obtuse angle (i.e. greater that 90° but less than 180°).

Area of a Triangle

There are **three** formulae used to find the area of a triangle. Lets consider each one in turn.

Formula 1

$$A = \frac{1}{2} \times b \times h$$

where b is the length of the **base and** h is the height of a right-angled triangle.

> Two right-angled triangles together create a rectangular of area of b x h. So the area of the triangle is simply half this value of (b x h)/2.

Formula 2

$$A = \frac{1}{2}(product\ of\ any\ two\ sides) \times \sin\theta$$

where θ is the angle between the known sides.

Heron's Formula 3

$$A = \sqrt{s(s-a)(s-b)(s-c)}$$

where a, b and c are the length of the sides of the triangle and s the semi-perimeter given by $s = \frac{a+b+c}{2}$

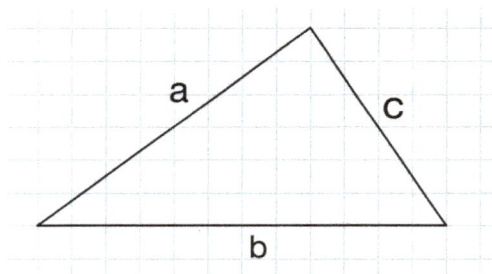

Sometimes, we know the lengths of the three sides of a triangle but not the perpendicular height nor an angle. When this happens, we should not use formula 1 or formula 2. Heron's formula enables us to find the area of a triangle from the lengths of the sides.

Examples

1.   Determine the area of the figure below.

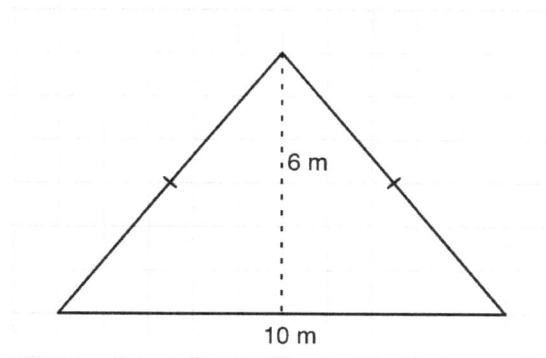

The triangle consists of two right-angled triangles side-by-side, with base of length 5 m and height 6 m. Thus, the area

$$A = \left(\frac{1}{2} \times 5m \times 6m\right) + \left(\frac{1}{2} \times 5m \times 6m\right)$$

$$= \frac{1}{2} \times 10m \times 6m$$

$$= \frac{1}{2} \times 60m^2$$

$$= 30m^2$$

2.   In the figure below, AB = 6cm, AC = 13cm and $\hat{A}$ = 50°. Calculate the area of the triangle.

$$A = \frac{1}{2}(product\ of\ any\ two\ sides) \times \sin\theta$$

$$= \frac{1}{2} \times (13cm \times 6cm) \times \sin 50^o$$

$$= \frac{1}{2} \times 78cm^2 \times 0.766$$

$$= \frac{1}{2} \times 59.75cm^2 = 29.875cm^2$$

$$\cong 29.9cm^2$$

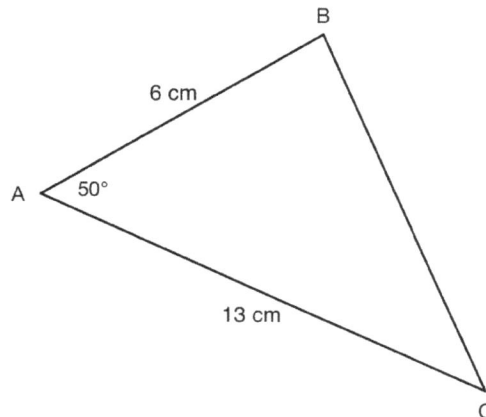

3.   Determine the area of the figure below.

$$s = \frac{a+b+c}{2} = \frac{3cm+4cm+5cm}{2} = \frac{12cm}{2} = 6cm$$

$$A = \sqrt{s(s-a)(s-b)(s-c)}$$
$$= \sqrt{6cm(6cm-3cm)(6cm-4cm)(6cm-5cm)}$$
$$= \sqrt{6cm(3cm)(2cm)(1cm)}$$
$$= \sqrt{36cm^4} = 6cm^2$$

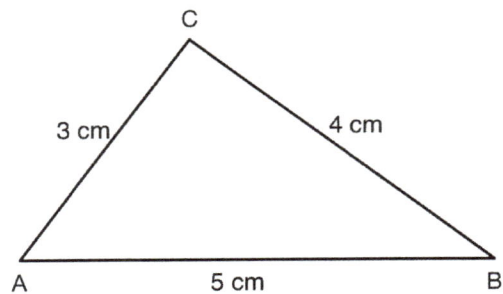

Area of a circle

The formula for the area of a circle is $A = \pi r^2$, where r is the radius of the circle.

Examples

1.   The diagram below, not drawn to scale, shows two circles with centre O and radii of 5cm and 7 cm. Calculate,

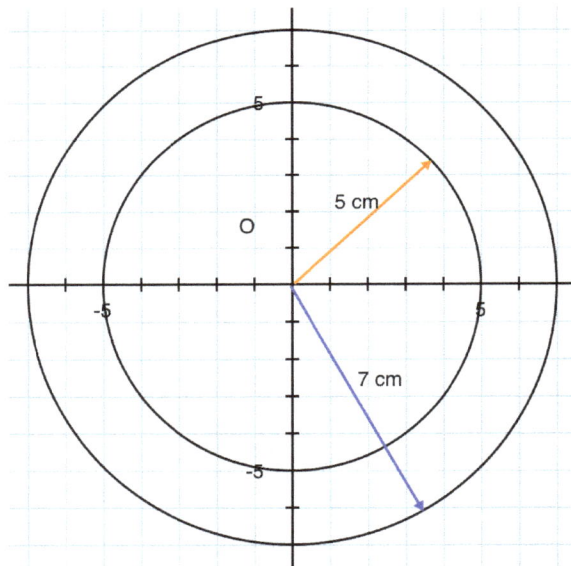

(a)   The area of the small circle.

$$A = \pi r^2 = 3.14 \times (5cm)^2 = 3.14 \times 25cm^2 = 78.5cm^2$$

(b)   The area of the bigger circle.

$$A = \pi r^2 = 3.14 \times (7cm)^2 = 3.14 \times 49cm^2 = 153.86cm^2 \cong 153.9cm^2$$

(c)   The area between the circles.

$$A = 153.9cm^2 - 78.5cm^2 = 75.4cm^2$$

2.   The diagram below, not drawn to scale, shows a sector with centre O and a square OABC. The radius of the sector is 4cm. Calculate,

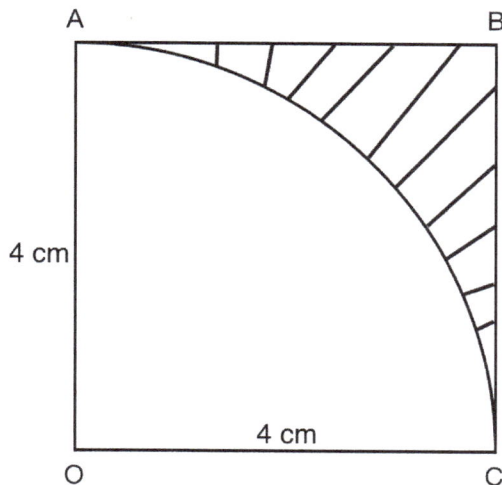

To find the area of the shaded region, we need to find the area of the sector OAC. Then subtract it from the area of the square.

(a)   The area of the sector.

$$A = \left(\frac{\theta}{360}\right) \times \pi r^2$$

$$= \left(\frac{90}{360}\right) \times 3.14 \times (4cm)^2$$

$$= \left(\frac{1}{4}\right) \times 50.24cm^2 = 12.56cm^2$$

(b)   The area of the square OABC.

$$A = l \times l = 4cm \times 4cm = 16cm^2$$

(c)   The area of the shaded region.

Area of the shaded region $= 16cm^2 - 12.56cm^2 = 3.44cm^2$

3.   The diagram below, not drawn to scale, shows a triangle OMN and a circle with O. The radius of the circle 4 cm. Calculate,

(a)   The arc length of MN.

$$l = \frac{\theta}{360} \times 2\pi r$$

$$= \frac{25}{360} \times 2 \times 3.14 \times 4 cm$$

$$= \frac{25}{360} \times 25.12 cm$$

$$= \frac{628 cm}{360}$$

$$= 1.74 cm$$

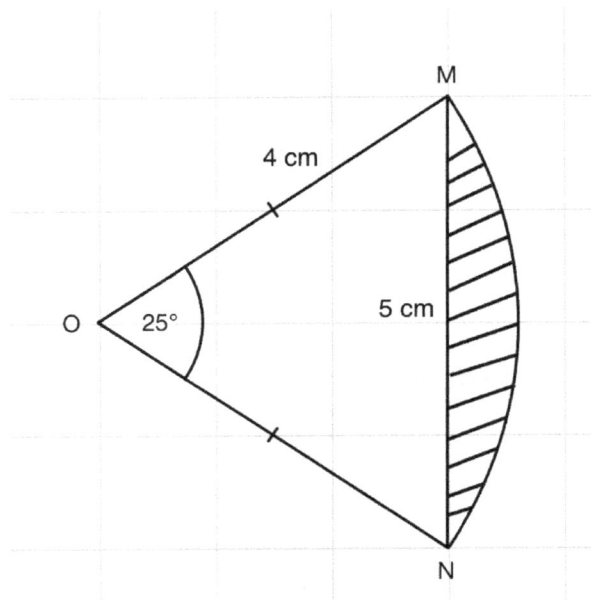

(b)   The area of the triangle OMN.

$$A = \frac{1}{2}(product\ of\ any\ two\ sides) \times \sin\theta$$

$$= \frac{1}{2} \times 4cm \times 4cm \times \sin 25$$

$$= \frac{1}{2} \times 16 cm^2 \times \sin 25$$

$$= 8 cm^2 \times 0.42 = 3.36 cm^2$$

(c)   The area of the sector OMN.

$$A = \frac{\theta}{360} \times \pi r^2$$

$$= \frac{25}{360} \times 3.14 \times (4cm)^2$$

$$= \frac{25}{360} \times 3.14 \times 16 cm^2$$

$$= 0.069 \times 50.24 cm^2 = 3.5 cm^2$$

(d)   The area of the shaded region.

Area of the shaded region $= 3.5 cm^2 - 3.3 cm^2 = 0.2 cm^2$

Combination of figures

To solve questions on figures that are the combination of shapes, identify the shapes, solve for each shape and add.

Examples

1.   In the diagram not drawn to scale, shows a triangle EDC and a Square ABCE.

(a)   Determine its perimeter.

$perimeter = AB + BC + CD + DE + EA$

$= 6cm + 6cm + 5cm + 5cm + 6cm = 28cm$

(b)   Determine its area.

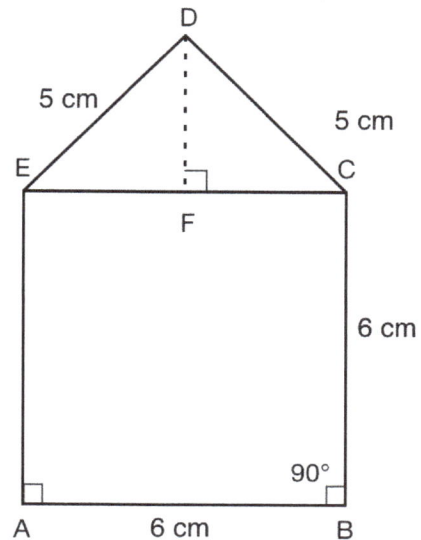

The figure can be broken down into two plane figures. Namely, a square and triangle as shown below.

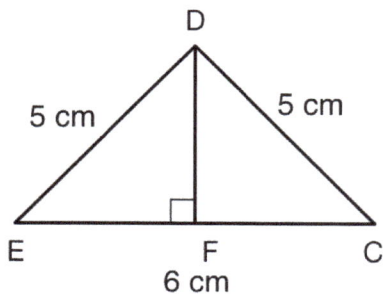

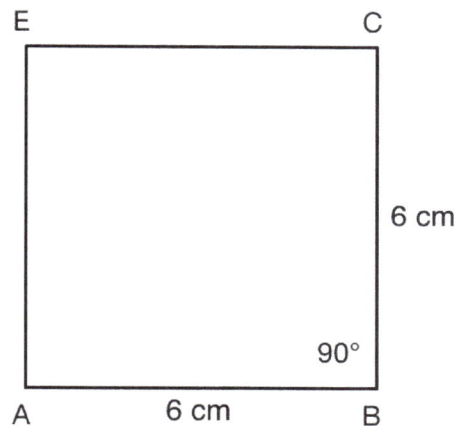

Note that this triangle DEC is an isosceles triangle. When we project the line DF, it cuts the base of the triangle in half giving us two right-angled triangles which are DEF and CDF. The length EF of the triangle DEF is 3 cm.

To find the area of the triangle, we need to find the perpendicular height FD of the triangle DEF using the Pythagoras theorem as follows.

$$EF^2 + FD^2 = DE^2$$
$$3^2 + FD^2 = 5^2$$
$$9 + FD^2 = 25$$
$$FD^2 = 25 - 9$$
$$FD^2 = 16$$
$$FD = \sqrt{16} = 4cm$$

Therefore, the area of the triangle DEF is $\frac{1}{2} \times 3cm \times 4cm = 6cm^2$ (or use Heron's formula) and area of triangle DEC is $6cm^2 + 6cm^2 = 12cm^2$

Now the area of square ABCE is $6cm \times 6cm = 36\ cm^2$

Finally, the area of the shape ABCDE is $= 36\ cm^2 + 12\ cm^2 = 48\ cm^2$

2.   Determine the area and perimeter of the figure below.

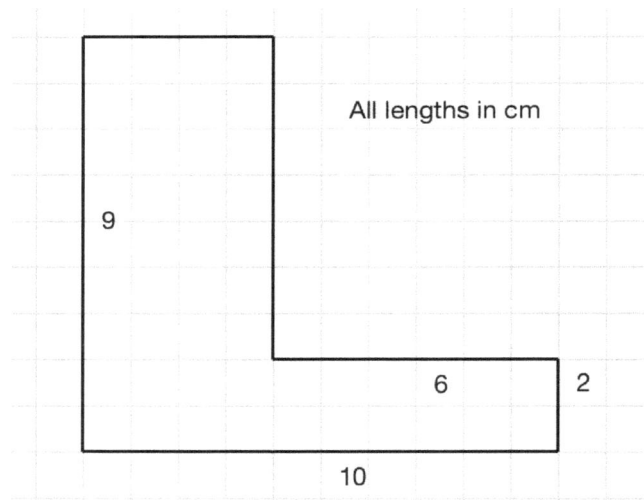

All lengths in cm

9

6   2

10

To find the area of the figure above, we can split it into two rectangles with the missing dimensions.

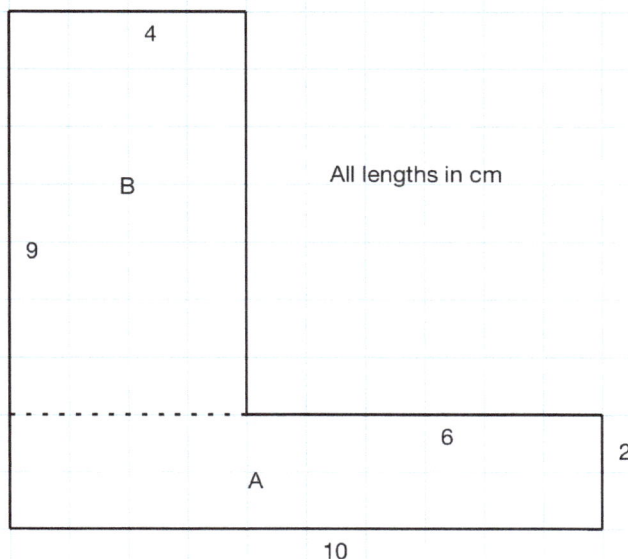

$$perimeter = 10cm + 2cm + 6cm + 7cm + 4cm + 9cm = 38cm$$

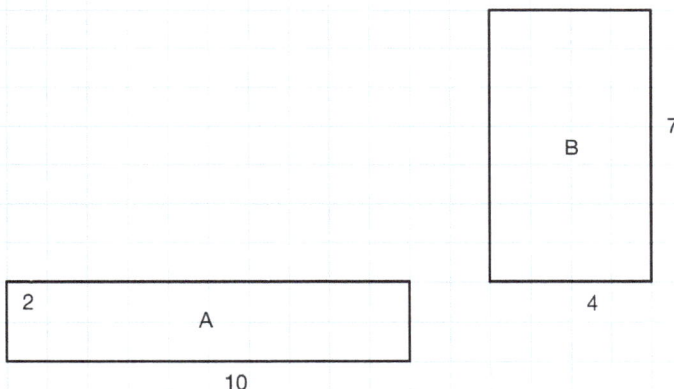

$Area\ of\ Rectangle\ A = 10cm \times 2cm = 20\ cm^2$

$Area\ of\ Rectangle\ B = 7cm \times 4cm = 28\ cm^2$

$Area\ of\ the\ figure = 28\ cm^2 + 20\ cm^2 = 48\ cm^2$

Density

Density of a solid is defined as its mass per unit volume.

- The formula for density is $D = \dfrac{mass}{volume}$

'per unit' means to divide.

Capacity

The **capacity** of a container is the amount of space (i.e. volume) that a liquid takes up within the container. The SI unit of capacity is the litre (l).

- A **litre** is the volume of a cube of side 10cm. That is, 1000cm^3 = 1 litre.
- 1 litre =1000ml (1ml = 1cm^3)
- 1000 litres = 1m^3
- 1000 litres = 1 kilolitre (kl)

Surface area and volume

Surface area is the area of the surface that bounds a solid.

Volume is the measure of the amount of space occupied by a three-dimensional object.

- The general formula for the **volume** is
$V = area\ of\ base(or\ a\ face) \times height\ (or\ length)$

> The volume is only found with three dimensional objects.

Cylinder

For the cylinder shown

- *Curved surface area of cylinder = circumference of base × height*
  $= 2\pi rh$

- *Surface area of cylinder* $= 2\pi rh + 2\pi r^2$ (including base and top of the cylinder)

- *Volume of cylinder = area of base × height*

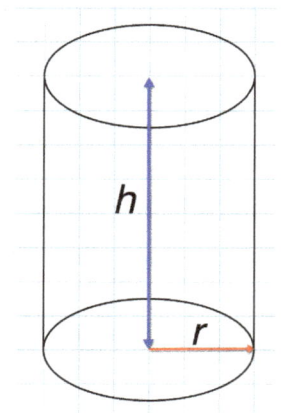

Examples

1.   The diagram below, not drawn to scale, shows a cylinder with height 10cm and radius 5cm. Calculate,

a)   The curved surface area of the cylinder.

Curved surface area of cylinder = circumference of base × height

$= 2\pi rh$

$= 2 \times 3.14 \times 5cm \times 10cm = 314 cm^2$

b)   The total area of the cylinder if both the base and top are closed.

Surface area of cylinder $= 2\pi rh + 2\pi r^2$

$= (2 \times 3.14 \times 5cm \times 10cm) + (2 \times 3.14 \times (5cm)^2)$

$= 314 cm^2 + 157 cm^2 = 471 cm^2$

c)   The volume of the cylinder.

Volume of cylinder = area of base × height

$= \pi r^2 \times height$

$= 3.14 \times (5cm)^2 \times 10cm$

$= 3.14 \times 25 \ cm^2 \times 10cm = 785 cm^3$

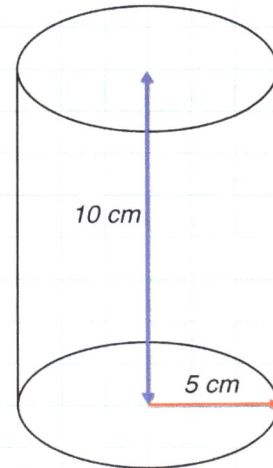

> The base of a cylinder is a circle.

d)   The capacity in litres of the cylinder.

$Capacity = \dfrac{785 cm^3}{1000 \ cm^3} = 0.785 \ litres$

> 1 litre = 1000 cm³

2.   Two litres of water is poured into a cylindrical bucket. If the base area of the bucket is 100 cm², what is the height of the water in the bucket?

1 litre = 1000 cm³
2 litres = 2 x 1000 = 2000 cm³
Volume of the water in the bucket is 2000 cm³.

Volume of cylinder = area of base × height

$2000 cm^3 = 100 cm^2 \times height$

$\dfrac{2000 cm^3}{100 cm^2} = height$

$20 cm = height$

Cone

A pyramid with a circular base is called a cone. For a cone with base radius *r*, slant height *s* and height *h*,

- *Curved surface area of cone = πrs*

- *Area of curved surface and base = πrs + πr²*

- *Volume of cone = $\dfrac{1}{3}\pi r^2 h$*

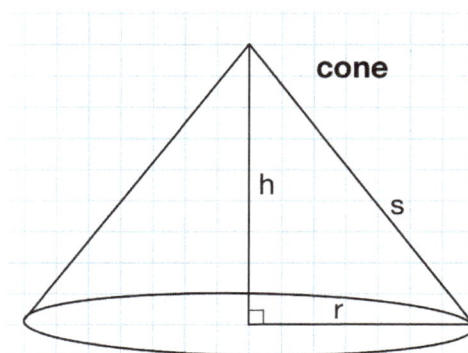

Example

A cone has a base of radius 6cm, slanting height of 15cm and a height of 10cm. Calculate,

a) The curved surface area of the cone.

Curved surface area of cone = πrs

$= 3.14 \times 6cm \times 15cm = 282.6 cm^2$

b) Area of the base.

Area of base $= 3.14 \times (6cm)^2 = 3.14 \times 36 cm^2 = 113.04\ cm^2$

c) The volume of the cone.

Volume of cone $= \dfrac{1}{3}\pi r^2 h$

$= \dfrac{1}{3} \times 3.14 \times (6cm)^2 \times 10cm = 376.8 cm^3$

Prism

A prism is a solid having a pair of parallel, congruent polygons as bases and parallelograms as sides. A prism is a called a right prism if the faces are perpendicular to the bases.

Lateral surface area of a right prism = perimeter of base × height
Surface area of a right prism including both bases = perimeter of base × height + 2(area of base)
Volume of a right prism = area of base × height

Examples

1.   The diagram below, not drawn to scale, shows a cuboid of length 6 cm, with 5 cm and height 4 cm.

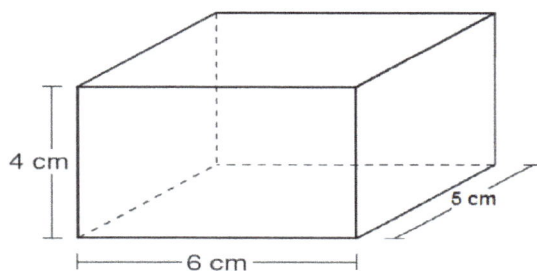

A cuboid has 6 faces. To find the total surface area, we need to add the area of the six faces.

a) The volume, in cm³, of the cuboid.

$V = area\ of\ base \times height$
$= 6cm \times 5cm \times 4cm = 120cm^3$

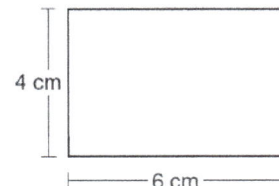

b) The total surface area, in cm², of the cuboid.

$Area = 6cm \times 4cm = 24cm^2$
$Total\ area = 2 \times 24cm^2 = 48cm^2$

$Area = 4cm \times 5cm = 20cm^2$
$Total\ area = 2 \times 20cm^2 = 40cm^2$

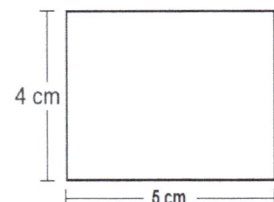

$Area = 6cm \times 5cm = 30cm^2$
$Total\ area = 2 \times 30cm^2 = 60cm^2$

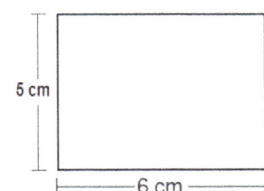

$Total\ surface\ area = 48cm^2 + 40cm^2 + 60cm^2 = 148cm^2$

2.   Pine Hill Dairy sells juice in boxes in the shape of a cuboid with internal dimensions 5cm by 4cm by 12.5 cm as shown below.

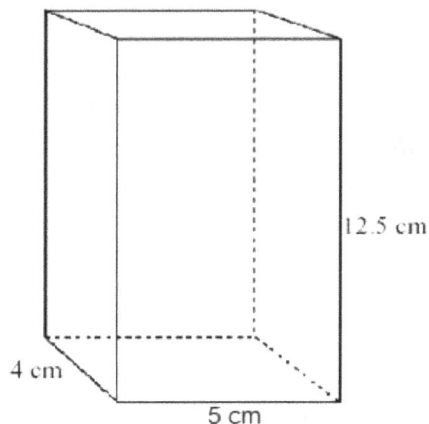

Calculate,

a) The volume, in cm³, of the box.

$V = length \times width \times height$

$= 5cm \times 4cm \times 12.5cm = 250cm^3$

b)   The capacity of the juice in litres.

$1 \ litre = 1000ml = 1000cm^3$

$1 \ ml = 1cm^3$

$Number \ of \ litres = \dfrac{250}{1000} = 0.25l$

c)   To make a smoothie requires 1 litre of juice. How many boxes are needed to make the smoothie?

$Number \ of \ boxes = \dfrac{1l}{0.25l} = 4$

d)   Suppose a smoothie has the same volume of the box. If the smoothie is poured into a cylindrical glass of internal diameter 7cm, what is the height of juice in the glass? Give your answer to 2 decimal places. Use $\pi = \dfrac{22}{7}$.

$Volume\ of\ smoothie = 250cm^3 = 250ml$

$Volume\ of\ juice = \dfrac{22}{7} \times r^2 \times h$

$250cm^3 = \dfrac{22}{7} \times (3.5cm)^2 \times h$

$250cm^3 = \dfrac{22}{7} \times \dfrac{12.25cm^2}{1} \times \dfrac{h}{1}$

$250cm^3 = 38.5cm^2 \times h$

$\dfrac{250cm^3}{38.5cm^2} = h$

$6.49cm = h$

The cylindrical glass volume =
area of circle x height

3.   The diagram below, not drawn to scale, shows a metal block. The cross section of the block is a trapezium, ABCD, with parallel sides AB and DC and AD, the perpendicular distance between the sides.

AB = 7 cm, BC = 6.5 cm, AD = 6 cm and DC = 5cm

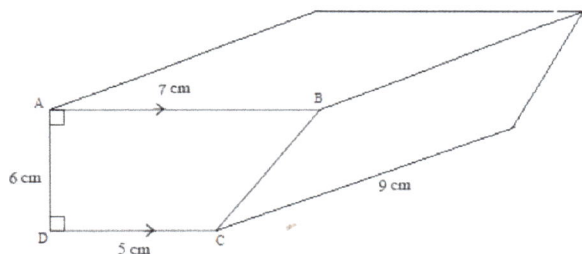

a) Calculate the area of the area trapezium ABCD.

$A = \left(\dfrac{a+b}{2} \right) \times h$

$= \left(\dfrac{7cm + 5cm}{2} \right) \times 6cm$

$= \left(\dfrac{12cm}{2} \right) \times 6cm$

$= 6cm \times 6cm = 36cm^2$

b) Calculate the volume of the metal block.

$Volume = Area\ of\ cross\ section\ x\ breadth\ of\ block$

$Volume = 36cm^2 \times 9cm = 324cm^3$

c) The block of metal has a mass of 1.5 kg. Calculate, in grams, the mass of one cubic centimetre of the metal.

$$Density = \frac{mass}{volume} = \frac{1500g}{324cm^3} = 4.6 \ gcm^{-3}$$

> The mass of one cubic centimetre of the metal means the density.
> 1kg = 1000 g
> 1.5kg = 1500g

3.   The diagram below, not drawn to scale, shows the cross section of a prism in the shape of a sector of a circle, centre O, and diameter of 10cm. The angle at the centre is 180°. Calculate,

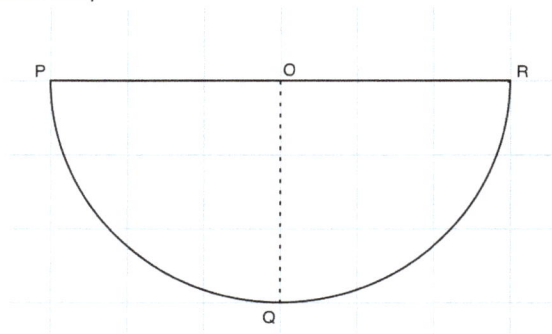

a) The length of the arc PQR.

$$length \ of \ arc = \frac{180}{360} \times 2 \times 3.14 \times 5cm$$

$$= \frac{1}{2} \times 34cm = 17cm$$

b) The perimeter of sector OPQR.
$$perimeter = PQR + ROP$$
$$= 17cm + 10cm = 27cm$$

c) The area of the sector OPQR.

$$Area \ of \ OPQR = \frac{\theta}{360} \times \pi r^2$$

$$= \frac{180}{360} \times 3.14 \times (5cm)^2$$

$$= \frac{1}{2} \times 3.14 \times 25cm^2 = \frac{1}{2} \times 78.5cm^2 = 39.3cm^2$$

d) The volume of the prism, if the prism is 10cm long and is a solid made of metal.

$Volume = area \ of \ OPQR \times length$

$= 39.3cm^2 \times 10cm = 393cm^3$

e) The mass of the prism, to the nearest kg, given that 1 cm^3 of metal has a mass of 7.3kg.

1cm^3 = 7.3 kg
393cm^3 = 7.3 x 393 = 2751 kg
Mass = 2751 kg

Right pyramid

A right pyramid is a pyramid where the vertex is directly above the centre of the base.

$Lateral \ surface \ area = \frac{1}{2} \times perimeter \ of \ base \times slanted \ height$

$Total \ Surface \ area = \frac{1}{2} \times perimeter \ of \ base \times slanted \ height + area \ of \ base$

$Volume \ of \ a \ right \ pyramid = \frac{1}{3} \times area \ of \ base \times height$

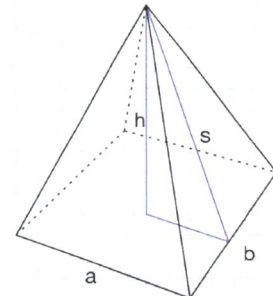

Example

1.   A right pyramid has a rectangular base of 8cm by 7cm and a slanting height of 12cm and height of 11cm. Determine:

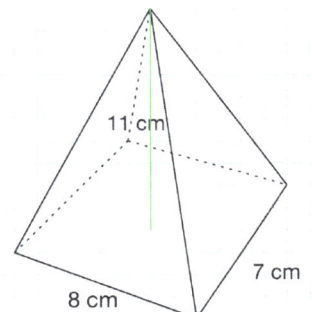

a) The lateral surface area.

$perimeter \ of \ base = 2(8cm) + 2(7cm) = 16cm + 14cm = 30cm$

Lateral surface area of a right pyramid $= \dfrac{1}{2} \times$ *perimeter of base* $\times$ *slanted height*

$= \dfrac{1}{2} \times 30cm \times 12cm = 180cm^2$

b) The total surface area.

Total surface area $= \left(\dfrac{1}{2} \times 30cm \times 12cm \right) + 56cm^2 = 180cm^2 + 56cm^2 = 236cm^2$

c) Its volume.

Volume of a right pyramid $= \dfrac{1}{3} \times$ *area of base* $\times$ *height*

$= \dfrac{1}{3} \times 56cm^2 \times 11cm = 205.3cm^3$

Sphere

A sphere is a three dimensional figure defined by the set of points that are a given distance from a defined point.

- *Surface area of sphere* $= 4\pi r^2$

- *Volume of sphere* $= \dfrac{4}{3}\pi r^3$

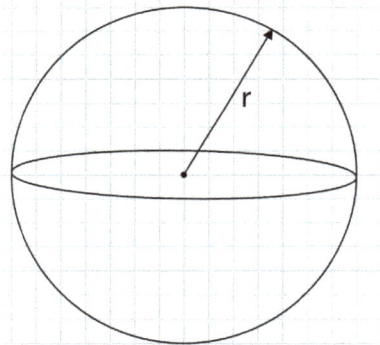

Example

Determine the surface area and volume of a sphere of radius 7cm.

Surface area of sphere $= 4\pi r^2$

$= 4 \times 3.14 \times (7cm)^2 = 4 \times \dfrac{22}{7} \times 49cm^2$

$= \dfrac{4312cm^2}{7} = 616cm^2$

$$Volume\ of\ sphere = \frac{4}{3}\pi r^3$$

$$= \frac{4}{3} \times \frac{22}{7} \times (7cm)^3 = \frac{88}{21} \times 343cm^3$$

$$= 1372cm^3$$

Scales

A scale is the ratio of the length in a drawing (i.e. a model) to the length of the original side.

Examples

1.   The scale of a map is 1: 1500000. The distance between two towns is 2.1cm. Calculate the actual distance on the ground between the two towns.
Scale of the map is 1: 1500,000. This ratio can be expressed as map distance to ground distance.

$$Actual\ distance = 2.1\ cm \times 1500000 = 3150000cm$$

> Now we can convert 3150,000 cm to km

1 m = 100 cm, therefore 3150000 ÷ 100 = 31500 m
1 km = 1000 m, therefore 31500 ÷ 1000 = 31.5 km is the ground distance.

2.   The scale on a map is 1:150,000. The distance on the map between two villages is 5.1cm. Calculate,

(a) The actual distance, in km, between the two villages.

$$Actual\ distance = 5.1cm \times 1500,000 = 7650000cm$$

1 m = 100 cm, therefore 7650000 ÷ 100 = 76500 m
1 km = 1000 m, therefore 76500 ÷ 1000 = 76.5 km is the ground distance.

(b) The maximum possible distance, in km, between the two villages.

Maximum distance = actual distance + **error**

To find the error, we look at the place value 5 which is tenth i.e. $\left(\frac{1}{10} = 0.1\right)$

$$error = \frac{0.1}{2} = 0.05$$

Hence, the Maximum distance $= 76.5 + 0.05 = 76.55\,km$

(c) The distance on the map between two cities that are actually 71.3 km apart on the ground.

1 km = 1000 m
1 m = 100 cm
1 km = 1000 x 100 = 100,000 cm
Therefore, 71.3 km = 71.3 x 100,000 = 7130000 cm
Map scale is 1: 1500,000

$$Distance\ on\ the\ map = \frac{7130000cm}{1500000} = 4.75\ cm$$

3.   A cylindrical tank has a base diameter of 2.8 m and a height of 5m. A model of the tank has a height of 10cm. Calculate:

(a) The scale used to make the model.

model height : tank height
10 cm : 5 m
1 m = 100 cm
5 cm : 5 x 100 = 500 cm

Reduce to the lowest terms
$$\frac{10cm}{10cm} : \frac{500cm}{10cm}$$
$1 : 50$

1 unit on model = 50 units on actual tank.

(b) The volume of the model.

$$tank\ radius = \frac{2.8\ cm}{2} = 1.4m = 140cm$$

$$model\ radius = \frac{140\ cm}{50} = 2.8cm$$

Recall scale is 1 : 50

$$Volume = \pi r^2 h$$

$$= \frac{22}{7} \times (2.8cm)^2 \times 5cm \quad 3cm$$

$$= 123.2 \ cm^3$$

4.   The diagram below is a map of a golf course drawn on a grid of 1cm squares. The scale of the map is 1:30000.

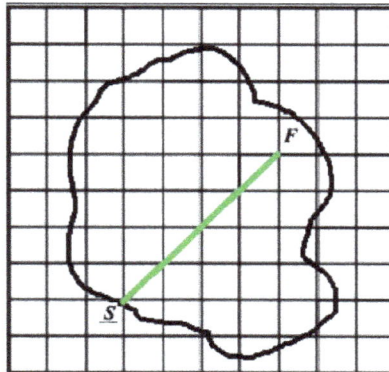

Using the map of the golf course calculate,

(a) The distance, to the nearest cm, from S to F.

Get your ruler and measure from S to F. The distance is 4.3 cm.

(b) The distance, in metres, from S to F on the actual golf course.

$$Scale = 1 : 30000$$
$$4.3cm \times 30000 = 129000cm$$
$$1m = 100cm$$

Converting to metres
$$\frac{129000}{100} = 1290m$$

(c) Stefan drove a golf cart from S to F in 150 seconds. Determine the golf cart's average speed in m/s and in km/h

$$speed = \frac{distance}{time} = \frac{1290m}{150 \ s} = 8.6 \ m/s$$

$$speed = \frac{1.29 \ km}{0.04} = 32.25 \ km/h$$

(d) The area on the ground represented by 1cm² on the map.

1 cm represents 30000 cm
30000 cm = 300 m
1 cm represents 300 m
1 cm² represents (300 m)²

$$(300m)^2 = 300m \times 300m = 90000m^2$$

(e) The actual area of the golf course, giving the answer in square metres

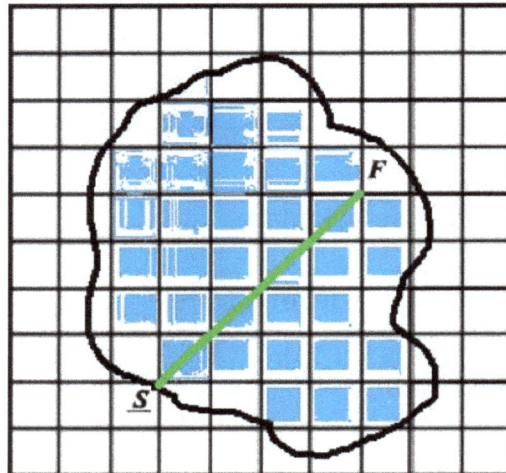

There are a total of 57 squares painted in blue and 33 squares not fully filled by the area of the golf course.

$$average = \frac{57 + 33}{2} = \frac{90}{2} = 45$$

The golf takes up approximately 26 squares.
$$area = 90000m^2 \times 45 = 4050000 \ m^2$$

Temperature

To change a temperature from degrees Celsius (^{0}C) to degrees Fahrenheit (^{0}F) use the formula

$$F = \frac{9C}{5} + 32$$

Similarly to change from Fahrenheit (^{0}F) to Celsius (^{0}C) use

$$C = \frac{5}{9}(F - 32)$$

If you remember one of these you should be able to work out the other.

Example

Convert 75F to degrees Celsius.

$$C = \frac{5}{9}\left(75^0 - 32\right) = \frac{5}{9} \times 43^0 = \frac{215^0}{9} = 23.89^0$$

Distance, time and speed

- Speed is the rate of change of distance with time.

- Average speed is the rate of change of total distance with total time.

- Velocity is the rate of change of distance with time in a specified direction.

- Acceleration is the rate of change of velocity with time.

If a cyclist travels 60 kilometres in 2 hours, we say his average speed is 30km/h (i.e. 30 km per hour). However, at some instants during his journey the speed would be less than 30km/h and at the other instants it would be greater.

$$average\ speed = \frac{total\ distance\ travelled}{total\ time\ taken}$$

Examples

1.   A bus leaves Bathsheba at 08:07 hrs and arrives at Bridgetown 13 km away at 08:45 hrs.

(a)   Find the time taken, in minutes, for the journey.

hrs	min
08	45
08	07
00	38

_ (subtraction)

$$Time\ taken\ =\ 38\ mins\ =\frac{38}{60}=0.63hrs$$

(b)   Find the average speed of the car in km/h.

$$average\ speed=\frac{13\ km}{0.63hrs}=20.6\ km/h$$

2.   Renaldo left Town A at 06:30 hours and travels to Town B in the same time zone.

(a)   He arrives at Town B at 1310 hours. How long did the journey take?

hrs	min
13	70
06⁷ 30	
06	40

(we cannot take away 30min from 10min, so we add 60 mins)

or equivalently

hrs	min
13¹²	70
06	30
06	40

(we cannot take away 30min from 10min, so we add 60 mins)

Time taken is 6 hrs and 40 mins or $6\frac{2}{3}hrs$

(b)   Renaldo travelled 310 kilometres. Calculate his average speed in km/h.

$$average\ speed=\frac{310km}{6\frac{2}{3}hrs}=46.5\ km\ h^{-1}$$

3.   The following is an extract from a bus schedule. The bus begins its journey at Belleview, travels to Chagville and ends its journey at St. Andrews.

Town	Arrive	Depart
Belleview	_____	6.40 am
Chagville	7.35 am	7.45 am
St. Andrews	8.00 am	_____

a)   How long did the bus spend at Chagville?

```
   hrs    min
   07     45
_  07     35
   00     10
```

Time spent at Chagville is 10 mins

b)   How long did the bus take to travel from Belleview to Chagville?

```
   hrs       min
   07 06     95      (added 60 mins)
_  06        40
   00        55
```

Time taken from Belleview to Chagnville = 55 mins or 0.92 hrs

c)   The bus travelled at an average speed of 54 km/hour from Belleview to Chagville. Calculate, in kilometres, the distance from Belleview to Chagville.

$$distance = 54\frac{km}{h} \times 0.92\ h = 49.7\ km$$

Travel graphs

Travel Graphs are classified two types of graphs:

- **Distance-Time graph** is the relationship between the distance on the y-axis and the time on the x-axis. The slope of the graph is the speed of the moving object.

If the car's speed has a direction then the speed becomes velocity. By definition, velocity is speed in a specific direction.

- **Speed-Time graph** is the relationship between the speed on the y-axis and the time on the x-axis. The slope of this type of graph is acceleration. The area under a speed time graph represents the distance travelled.

- **Velocity-Time graph** is the relationship between the velocity on the y-axis and the time on the x-axis. The slope of this type of graph is acceleration. The area under a velocity time graph represents the distance travelled.

Examples

1.   The diagram below represents the 5-hour journey of an athlete.

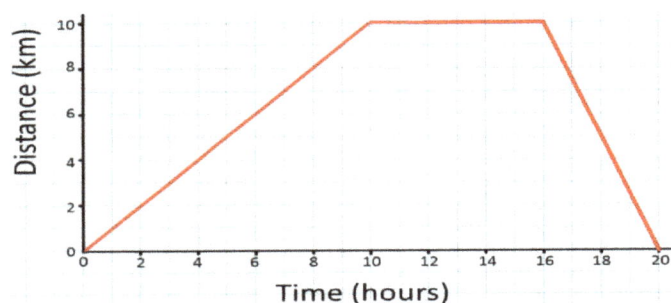

Always pay close attention to the labelling on the axes. Distance (km) is on the vertical axis and time is on the horizontal axis.

(a) What was the average speed during the first 10 hours?

$$average\ speed = \frac{10\ km}{10\ hrs} = 1\ km/h$$

(b) What did the athlete do between 10 and 16 hours after the start of the journey?

Between hours ten and sixty the athlete was resting.

(c) What was the average speed on the return journey?

The athlete resting time ended at 16 hrs. Now, athlete's return journey finished at 20 hours. Therefore, return journey took the athlete 4 hours. The distance of the return journey is 10 km.

$$average\ speed = \frac{10km}{4hrs} = 2.5\ km/h$$

2.   The speed-time graph below shows the movement of a motor car.

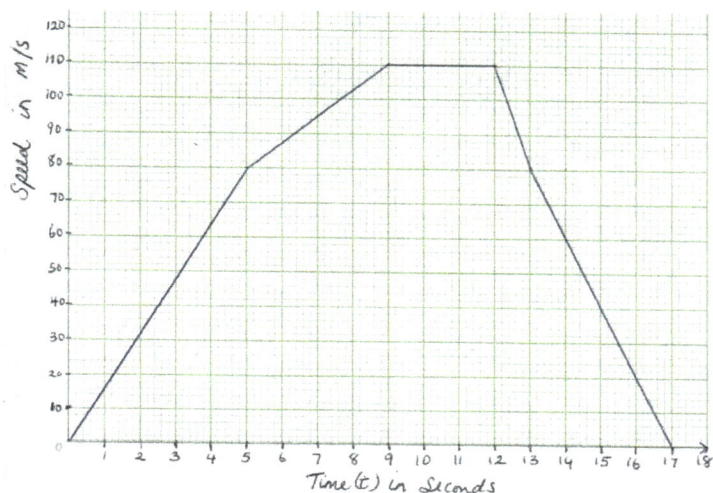

With speed on the vertical axis and time on the horizontal axis, the slope of the graph is acceleration.

Using the graph, calculate

(a) The acceleration of the motor during the first 5 seconds.

$$acceleration = \frac{change\ in\ velocity}{time} = \frac{final\ velocity - initial\ velocity}{time}$$

Speed = 80 m/s

$$acceleration = \frac{80\ m/s - 0\ m/s}{5\ s}$$

$$= \frac{80\ m/s}{5\ s} = 16 \left(\frac{\frac{m}{s}}{s} \right) = 16m/s^{-2}$$

$$\left(\frac{\frac{m}{s}}{s} \right) = \frac{m}{s} \div \frac{s}{1} = \frac{m}{s} \times \frac{1}{s} = \frac{m}{s^2} = ms^{-2}$$

(b) The distance traveled by the motor car.

Using the formula for the area of a trapezium will give the total distance travelled by the motor car since the area under the graph is the distance travelled.

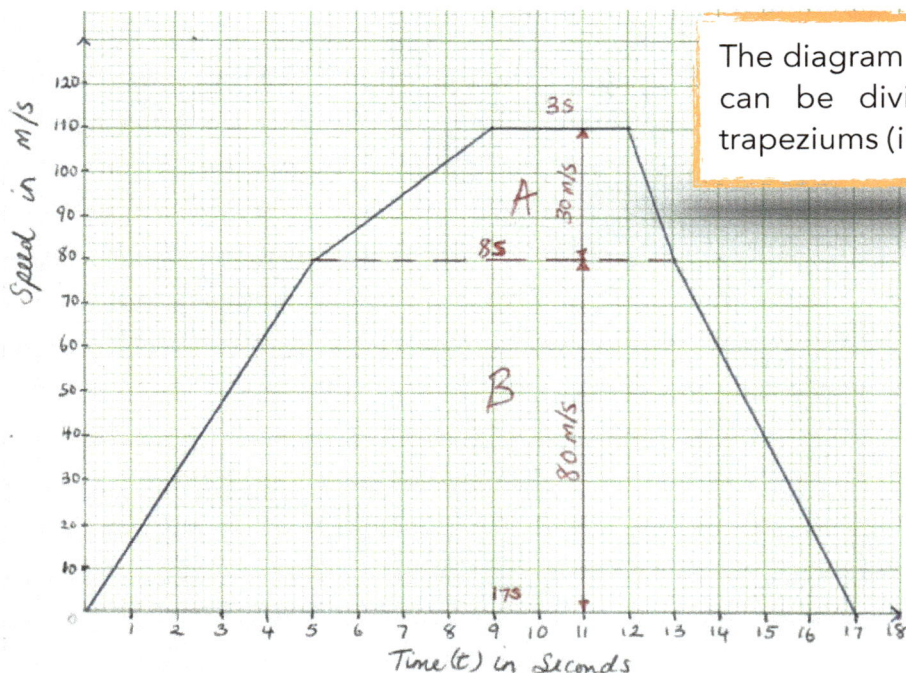

The diagram in the question can be divided into two trapeziums (i.e. A and B).

$trapezium\ area\ A = \left(\dfrac{8s+3s}{2}\right) \times 30m/s$

$= \dfrac{11s}{2} \times \dfrac{30m}{s} = \dfrac{330ms}{2s} = 165m$

$trapezium\ area\ B = \left(\dfrac{8s+17s}{2}\right) \times 80m/s$

$= \dfrac{25s}{2} \times \dfrac{80m}{s} = \dfrac{2000ms}{2s} = 1000m$

Hence, total distance = 1000m + 165m = 1165m.

3.   A cyclist travels at uniform velocities of 0 m/s for 3 seconds and then reaches 8 m/s in the next 2 seconds. He then undergoes a constant velocity for the next 3 seconds, and finally comes to rest with uniform deceleration in the next 2 seconds.

Sketch a velocity-time graph of the cyclist and use the graph to determine the,
(a) total time it took the cyclists to complete the journey.

Sketch of the velocity-time graph is shown below.

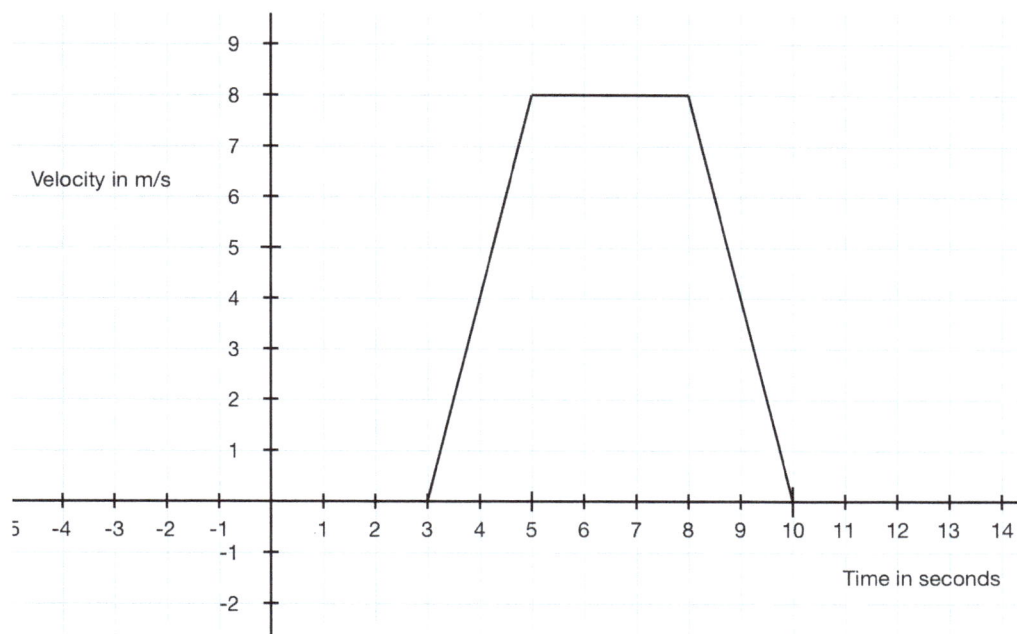

Total time take in 10 seconds.

(b) maximum velocity reached.

Maximum velocity is 8 m/s.

(c) rate of acceleration between 3 and 5 seconds.

Acceleration is a rate of change of velocity due to a specified time. Therefore, acceleration is the slope (i.e. gradient) of the graph. We can use $m = \frac{y_2 - y_1}{x_2 - x_1}$. The coordinates to substitute in this formula are (3,0) and (5,8).

$$m = \frac{y_2 - y_1}{x_2 - x_1} = \frac{8 - 0}{5 - 3} = \frac{8}{2} = 4 \ m/s^{-2}$$

$$\frac{m}{s} \times \frac{1}{s} = \frac{m}{s^2} = m/s^{-2}$$

or

$$acceleration = \frac{change \ in \ velocity}{time} = \frac{final \ velocity - initial \ velocity}{time \ taken}$$

$$= \frac{8 \ m/s - 0 \ m/s}{2 \ s} = \frac{8 \ m/s}{2 \ s} = 4 \ m/s^{-2}$$

(d) rate of acceleration between 5 and 8 seconds.

Since velocity is constant between 5 and 8 seconds acceleration is zero.
(e) rate of deceleration.

$$deceleration = \frac{initial\ velocity - final\ velocity}{time\ taken}$$

$$= \frac{8\ m/s - 10\ m/s}{2\ s} = \frac{-2\ m/s}{2\ s} = -1\ m/s^{-2}$$

> We can use any two of the formulae in part (c).

(f) total distance travelled.

Distance is the area under the graph. This is broken down into two sections. First section is the velocity of 0 m/s and 3 seconds.

$$displacement\ (ie\ distance) = 0\ m/s \times 3\ s = 0\ m$$

Second section is the plane surface we have in your graph which is a trapezium. Therefore, to find the distance traveled, we use the formula for the area of a trapezium.

$$= \left(\frac{a+b}{2}\right) \times h$$

$$a = 3, b = 7\ and\ h = 8$$

> $$s \times m/s = \frac{s}{1} \times \frac{m}{s} = m$$

$$A = \left(\frac{3+7}{2}\right) \times 12 = \frac{10\ s}{2} \times 12\ m/s = 60\ m$$

Questions
Measurement
Michael Boyce
Alexandra School
Barbados

> ‣ Video solutions to these questions via the App "**CTS Maths**" for your smartphone or tablet. Details on http://CaribbeanTeachersSeries.com

QUESTIONS

[1] The scale on a map is 1:250,000. The distance between two houses on the map is 3.8 cm. Calculate the actual distance on the ground between the two houses.

[2] The scale on a map is 1:150,000. The distance on the map between two campsites is 4.9 cm. Calculate the
(a) actual distance, in km, between the two campsites.
(b) maximum possible distance, in km, between the two campsites.
(c) distance on the map between two cities that are actually 6.4 km apart on the ground.

[3] A cylindrical tank has a base diameter of 1.4 m and a height of 4 m. A model of the tank has a height of 5 cm. Calculate,
(a) the scale used to make the model.

(b) the volume of the model.

[4] A car leaves Morris Town at 07:48 hrs and arrives at New Town 7.26 km away at 08:16 hrs. Calculate,
(a) the time taken in minutes for the journey.
(b) the average speed of the car in km/h.

[5] On Christmas day,
(a) A car left Jones Town at 13:30 hrs and reached Broome Town at 16:00 hrs. If it travelled at an average speed of 75 km/h, how far apart are the two towns?
(b) The car then left Broome Town at 17:20 hrs and arrived at Best Ville which is 120 km away at 21:05 hrs. Calculate the average speed for the journey from Jones Town to Best Ville.

[6] A flight leaves an airport in Barminca at 09:30 hrs local time and travels to Mesquay arriving at 13:30 hrs local time.   (NB : Mesquay is 1 hour ahead of Barminca)
(a) How long does it take to fly from Barminca to Mesquay?
(b) If Mesquay is 1089 km away from Barminca, at what speed did the flight travel? The return flight left Mesquay at 19:25 hrs local time and should have taken the same amount of time.   However, due to favorable wind conditions, it arrived 25 minutes earlier.
(c) What was the speed of the return flight?
(d) At what local time did the flight arrive in Mesquay?

[7] Joseph has two cylindrical containers. One has a base radius of 15 cm and a height of 28 cm. The second one has a base radius of 28 cm and a height of 15 cm. Which has the greater capacity?

[8] Cheese is sold in wedges in the shape of triangular prisms, the base of which is a right-angled triangle with perpendicular sides measuring 8 cm and 5 cm. The prisms have a length of 4 cm.   How many such wedges may be cut from a rectangular block 8 cm by 10 cm by 1 m?

[9] A rectangular water tank is 15 m long, 12 m wide and 8 m high.   The tank is filled with water. The water is then pumped into a cylindrical tank at a rate of 30 cubic meters per min so that it reaches a height of 8 m in the cylinder. Calculate:
   (a) The capacity of the water tank.
   (b) How long it takes, in hours, to pump all the water from the original tank.
   (c) The radius of the cylindrical tank. Use $\pi = \dfrac{22}{7}$.

[**10**] A cylinder has a base diameter of 21cm and a height of 15cm. Calculate,
(a) the area, in cm², of the base.

(b) the capacity, in litres, of the cylinder. Use $\pi = \dfrac{22}{7}$ and 1 litre = 1000cm³.

[**11**] A miniature statue made of steel is in the form of a rectangular pyramid with perpendicular height 9cm and base dimensions 8cm by 12cm. The statue is melted down and moulded into a cylinder with base radius 6cm. Calculate the height of the cylinder.

[**12**] A juice container is cylindrical in nature with a base radius of 7cm and a height of 21cm.
(a) Calculate the volume of the container.
(b) This container is used to pour juice into a bigger container of height 42cm. If 5 sets of juice must be poured using the smaller container to full the larger container, calculate the radius of the bigger container.

Use $\pi = \dfrac{22}{7}$ $V = \pi r h$

[**13**] A cube and a cuboid have the same height. Further, the volume of each is 729cm³. If the cuboid has a width of 18cm, find its length.

[**14**] A cylindrical barrel has a base diameter of 21cm and a height of 90cm. Calculate the capacity of the barrel. Water is poured into the barrel at a rate of 125cm per minute. How long will it take in hours to fill the barrel? Use $\pi = \dfrac{22}{7}$.

[**15**] The diagram below shows a right angled triangular based prism with QR=20cm, PQ=RS=8cm and PT=10cm. Calculate the volume of the prism.

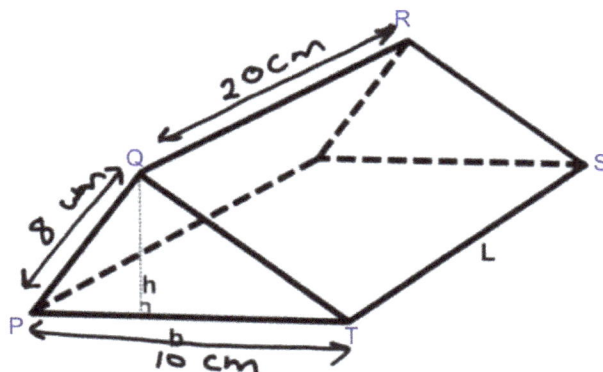

[16] A barbell in a gym is made from a cylindrical piece of iron of radius 20cm and 5cm thick. A hole 3cm in diameter is at the centre of the barrel and goes its entire thickness.
(a) Calculate the volume of the barbell.
(b) If 1m³ of iron has a mass of 7870kg, calculate the mass of the barbell.

[17] A storage bin is in the form of an inverted cone. The top of the bin has a radius of 10m . The slant height of the bin is 26m.
(a) Determine the perpendicular height of the storage bin.
(b) Hence, determine the capacity of the storage bin

N.B. $\pi = \dfrac{22}{7}$ $V = \dfrac{1}{3}AH$

[18] A feed bin is in the form of a square based pyramid with base side 20m and perpendicular height of 15m.
(a) Calculate the capacity of the bin.
(b) How long would a full bin of feed last a farmer whose chickens consume 50m³ of feed per day?

N.B. $V = \dfrac{1}{3}AH$

[19] The diagram below shows the cross section of a sardine tin. If the tin has a height of 2 cm, calculate its volume.

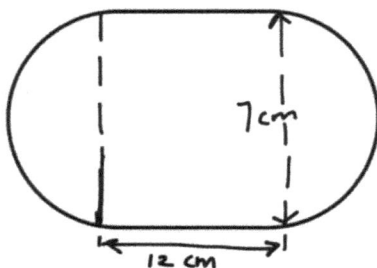

7cm

12 cm

[20] A bucket is in the shape of a frustum of a cone. The base of the bucket has a radius of 14 cm and the top has a radius of 21 cm. The height of the bucket is 10 cm. The frustum was made by removing a cone of height 7 cm from the bottom of the cone. Calculate the volume of the frustum. Use $\pi = \dfrac{22}{7}$.

Chapter 7
Geometry & Trigonometry
Jerome Stuart
Graydon Sealy School
Barbados

Geometry is all around us. It can be seen when we look at rainbows or in the lines of a beautifully designed sports car. The architectural designs of buildings also reflect our use of geometry. The Pentagon in the United States is designed in the shape of a pentagon while in the Caribbean, the wide variety of rooftops, windows and arches - with their myriad shapes and angles - all demonstrate the integral use of and need for geometry. This chapter will help you to understand the different theories in geometry and give you the methods and techniques for solving geometrical problems.

Key geometric terms

- A point is an exact position or location on a plane surface. It is usually represented by a dot that is encircled so that it can be easily identified. Points are often used in coordinate geometry. The diagram below shows the point A(1,2).

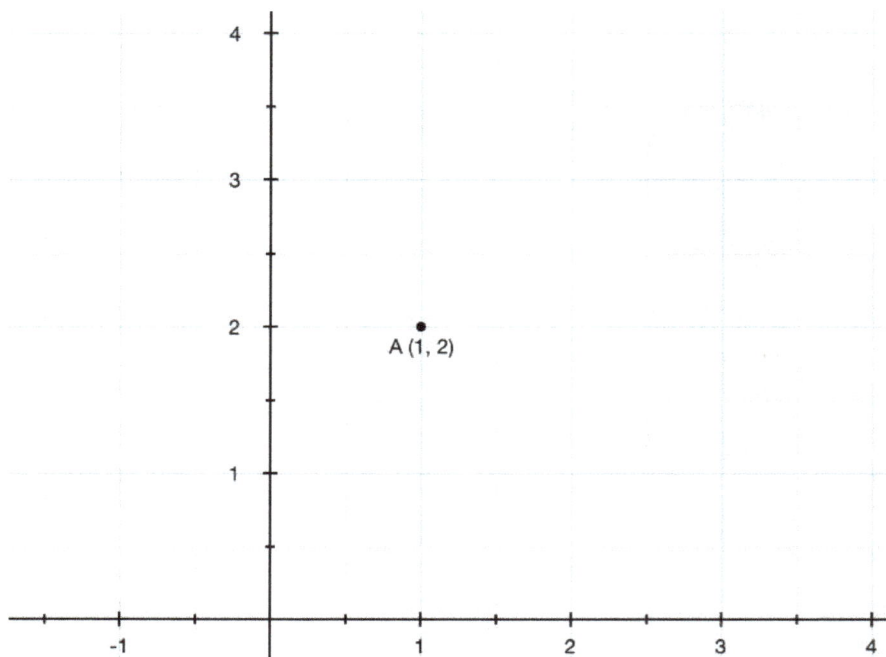

- A line can be either straight or curved. Line *a* is shown below.

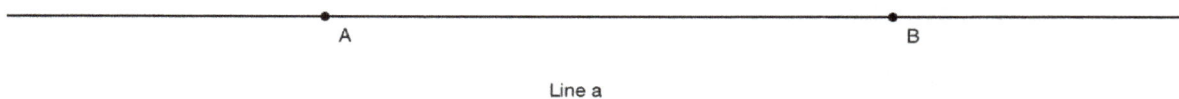

Line a

- A line segment is the part of a line consisting of two endpoints and all points in between them. The endpoints are indicated by capital letters. For example, AB above is a line segment with endpoints A and B.

- A ray is a straight line which starts at a point called the origin and extends in a particular direction to infinity. RS is an example of a ray, where R is the origin.

- Parallel lines are lines in the same plane that never intersect. They have the same gradient. Below, lines *b* and *c* are parallel.

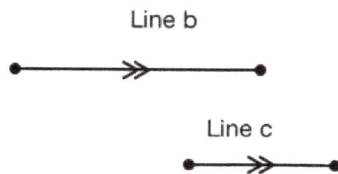

Line b

Line c

Parallel lines must never touch if the lines are extended.

- Intersecting lines are lines that cross each other.

- Perpendicular lines are lines that meet each other at right angles (90^0).

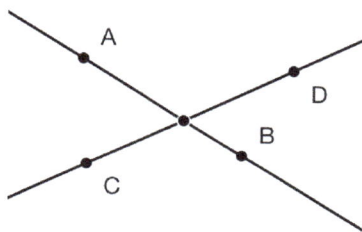

Intersecting lines

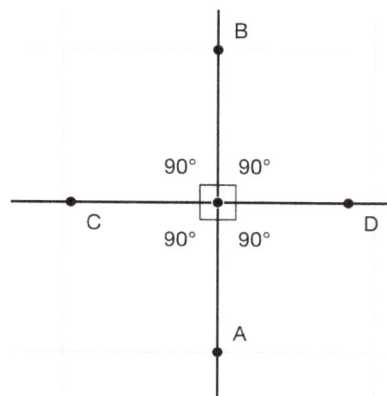

Perpendicular Lines

- A plane is a flat surface. For example, the screen of a laptop, the cover of a book or the surface of a tile.

- An angle is the amount of turn between two lines which have a common vertex. It is measured in degrees or radians. (π radians $= 180°$)

I.   An acute angle measures between $0°$ and $90°$.
II.   A right angle measures exactly $90°$.
III.   An obtuse angle measures between $90°$ and $180°$
IV.   A straight angle measures $180°$.
V.   A reflex angle measures more than $180°$ but less than $360°$.

Examples of different types of angles are shown below.

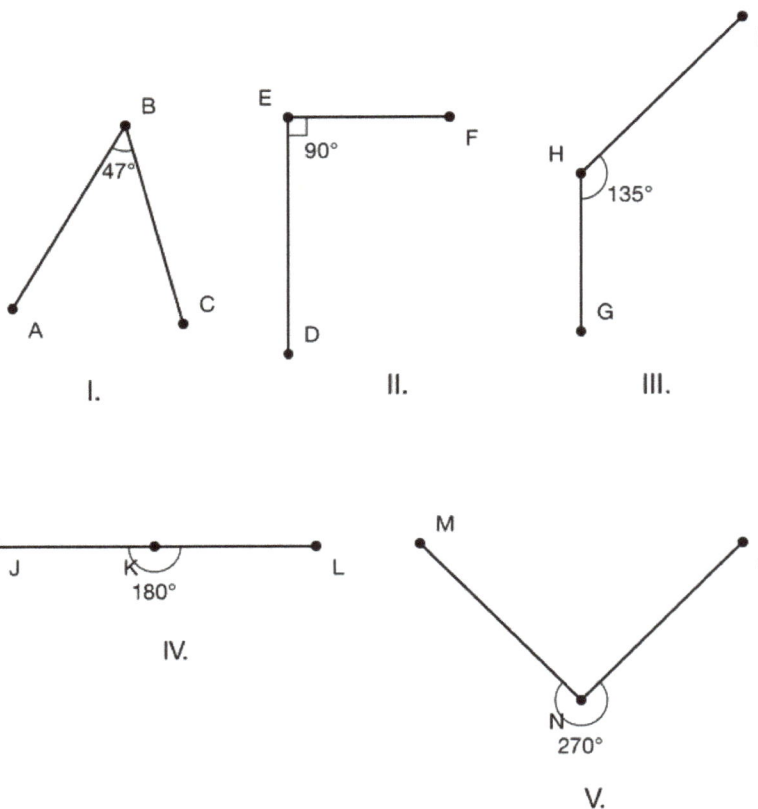

I.   II.   III.

IV.

V.

- A face is any surface of an object. Examples,

A circular face   Rectangular faces

- An edge is formed when two faces converge. For example, the red line above shows the edges of the cuboid.

- A vertex occurs where edges or lines meet at a point. For example, on the cuboid above, the red edges form vertices or corners where the edges meet.

Drawing and measuring angles and line segments

- To draw a line segment, use a metric ruler and compasses.
- To measure line segments, use a divider and a ruler.

- To draw angles, use a protractor and a ruler or straight edge.
- To measure angles in degrees, use a protractor.

Examples

1.   Construct a line segment EF of length 5 cm.

Step 1   Draw a straight line with length exceeding 5 cm.
Step 2   Indicate the location of endpoint E on the line.
Step 3   Open the compass to a distance of 5 cm.
Step 4   Place the compass point on point E and make an arc to intersect the line.
Step 5   Name the point of intersection F. Thus EF is the required line segment of length 5 cm.

2.   Measure the line segment PQ below.

Step 1   Open the divider and place one end on point P and the next end on point Q.
Step 2   Measure the distance that between the two open ends of the divider with a ruler. The distance is 7 cm.

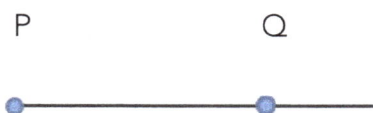

P   Q

3.   Draw an angle of 60°

Step 1   Draw arm BA.

B   A

Step 2   Fit the midpoint of the protractor's baseline on point B and line up the baseline of the protractor with the arm BA.
Step 3   Using the correct scale (the one which passes through A and starts at zero) make a point called C above the protractor in line with the $60°$
Step 4   Connect point C to point B. The angle constructed is a $60°$ angle.

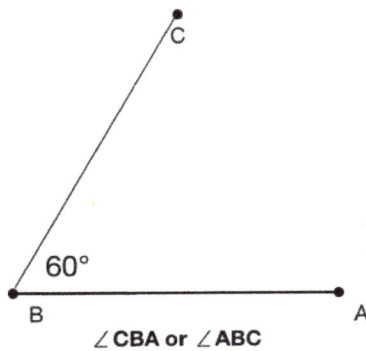

The angle is written as ∠CBA or ∠ABC

The letter B which is in the middle of ∠CBA is the vertex.

∠CBA or ∠ABC

4.   Find the size of the angle below.

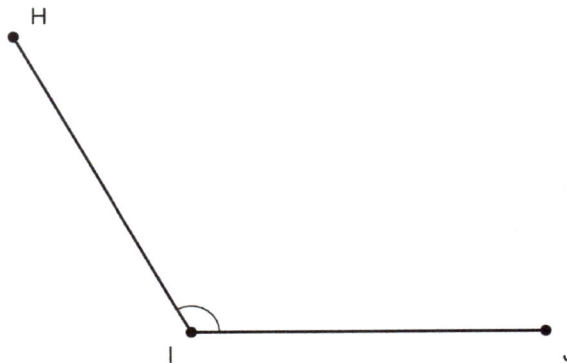

Step 1   Line up the base line of the protractor with either arm HI or IJ and make sure that the middle of the protractor is on the vertex I.

Step 2   Using the scale that starts at zero on the arm of the angle, determine where the next arm crosses the same scale of the protractor.

Step 3   Read off the measurement. The angle measured is 120˚.

To measure reflex angles measure either the acute or obtuse angle that is adjacent to it and then subtract from 360°.

Construction of lines, angles and polygons

- To construct angles we use a pair of compasses and a ruler. A constructed angle could be bisected if necessary to obtain an angle that is a half of the constructed angle.

- Polygons can be constructed using a pair of compasses, a ruler or straight edge and a protractor.

- Lines can be constructed using a set square and a ruler or a pair of compasses.

Examples

1.   Construct a 30° angle.

Step 1   Draw a line segment with end points P and Q.

Step 2   With the point of the compass at P, make an arc to intersect PQ at R. See diagram below.

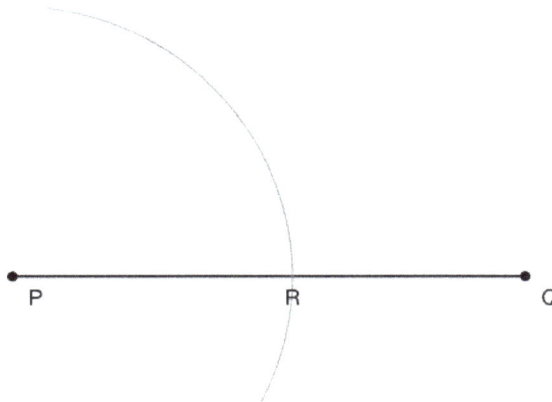

Step 3   Without changing the radius of the compass, place the point of the compass on point R and draw an arc. Let this arc also intersect the arc drawn in Step 2. Label this point of intersection S.

Step 4   Without changing the radius of the compass, place the point of the compass on point S and draw an arc to intersect the arc drawn in Step 3. Label this point of intersection T.

Step 5   Connect point P with the point T. The angle formed is a 30° angle. (See diagram below)

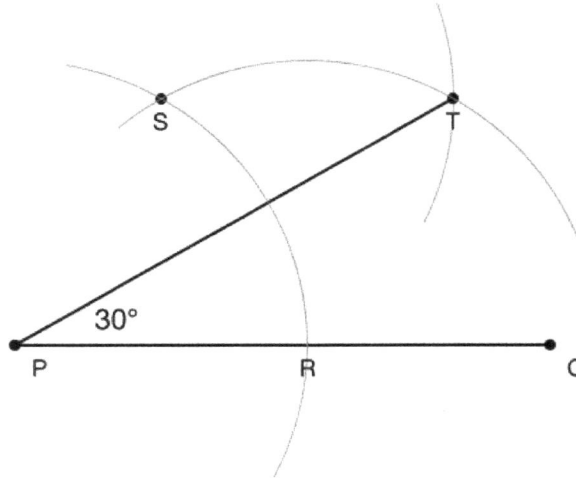

2.   Construct a 45° angle.

Step 1   Draw a line segment AB.

Step 2   Place the point of the compass at A and open the radius of the compass to more than half the length of AB and scribe two arcs: one above and one below AB.

Step 3   Without changing the radius of the compass place the point of the compass on B and scribe two arcs: one above and one below AB to intersect with the arcs made in the Step 2.

Step 4   Connect these two points of intersections with a line.

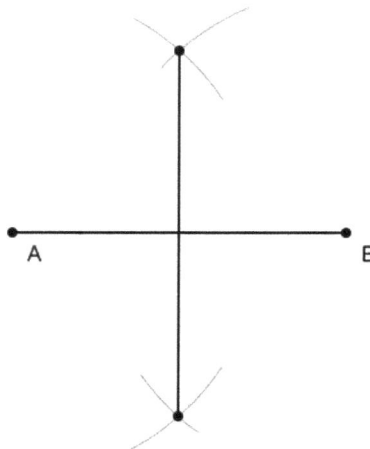

Step 5   Place the point of the compass on A and open the compass to the point where the perpendicular line crosses AB.

Step 6   Place the point of the compass on the point of intersection of AB and the perpendicular line and scribe an arc above AB to cross the perpendicular line. Name this point of intersection C.

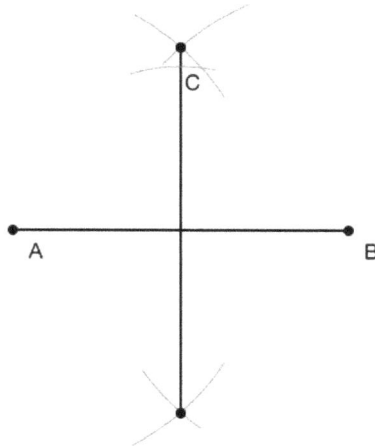

Step 7   Connect point A to point C and the angle formed is 45^0.

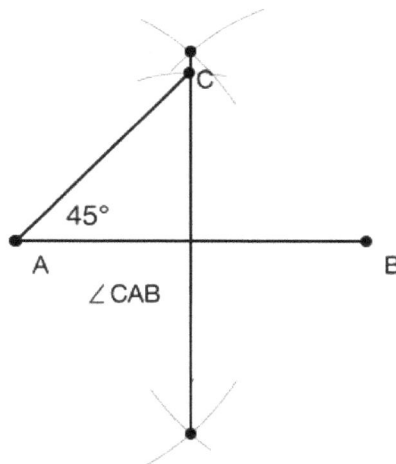

3.   Construct a 60° angle.

Angles in an equilateral triangle are all 60˚ in size.

Step 1   Draw a line segment with end points P and Q.

Step 2   With the point of the compass at P make an arc to intersect PQ at R.

Step 3   With the point of the compass at R, scribe an arc to intersect the arc drawn in step 2. Label this point of intersection S.

Step 4   Using a ruler draw a line from P to S. The angle formed is a 60° angle. (See <SPQ below). If an equilateral triangle is required, connect point R to point S and PRS is an equilateral triangle. (See ΔPRS below).

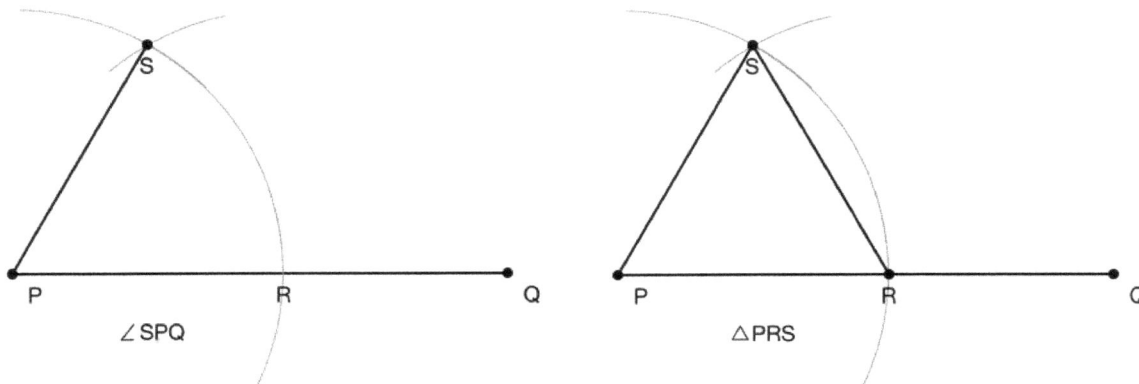

∠SPQ   ΔPRS

4.   Construct a 120° angle.

Supplementary angles sum to 180°. A 120^0 can be formed by constructing a 60^0 angle and then extending a line through the vertex so that the 120^0 is adjacent to the 60^0 angle.

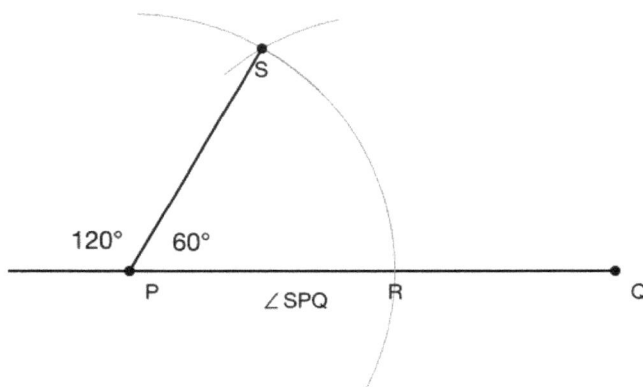

120°   60°

P   ∠SPQ   R   Q

5.   Constructing a 90° angle.
Step 1   Draw a line segment AB.
Step 2   Place the point of the compass on A and scribe an arc above line segment AB and below line segment AB.
Step 3   Place the point of the compass on B and scribe an arc above line segment AB to intersect the arc that was above AB. With the point of the compass still on point B, draw an arc below line segment AB to intersect the arc that was below AB.

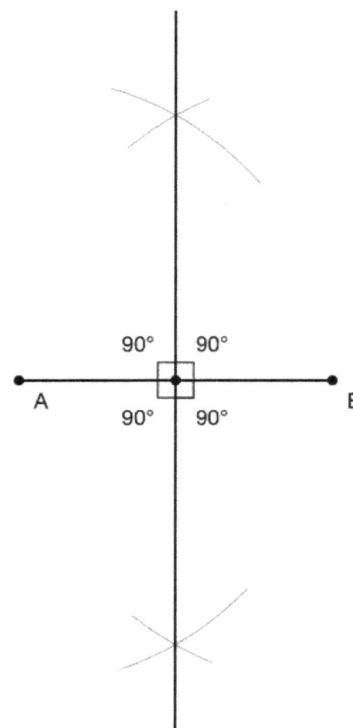

90°   90°

A   B

90°   90°

Step 4   Draw a line to connect the point of intersection above line AB to the point of intersection below line AB. The angles formed from the line intersecting AB are all 90^0 angles.

6.   Construct a line parallel to line segment AB.

Step 1   Draw line segment AB.
Step 2   Choose two points on AB and name them C and D.

Step 3   Place the compass point on C and make an arc above AB
Step 4   Without changing the radius of the compass, put the point of the compass on point D and scribe another arc above AB.

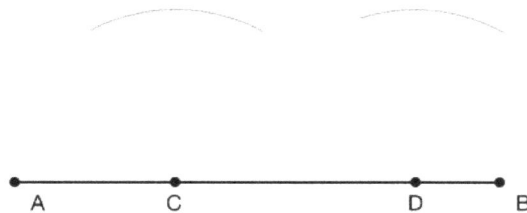

Step 5   Draw a line segment tangent to the two arcs made in Step 4. Label the starting point of this line E and the end point F. EF is parallel to AB.

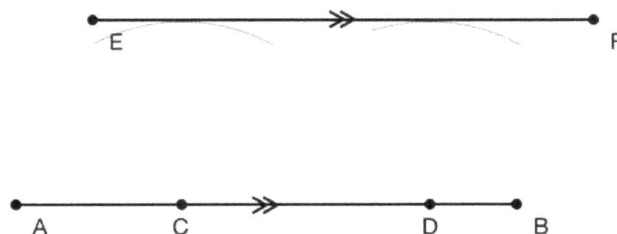

7.   Construct a line perpendicular to another line from a point not on the line.

Step 1   Draw a line segment AB and place a point C above AB.

Step 2　　Place the point of the compass on point C and scribe an arc to intersect AB at two places.

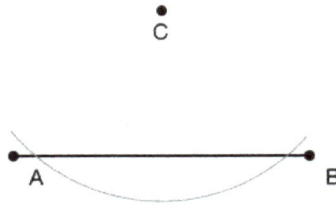

Step 3　　Open the compass wider and place the point of the compass on one point of intersection on AB and draw an arc. Without changing the radius of the compass, place the point of the compass on the other intersection of AB and draw another arc to intersect the arc just made.

Step 4　　Label this point of intersection of the two arcs D.

Step 5　　Connect point C and D.

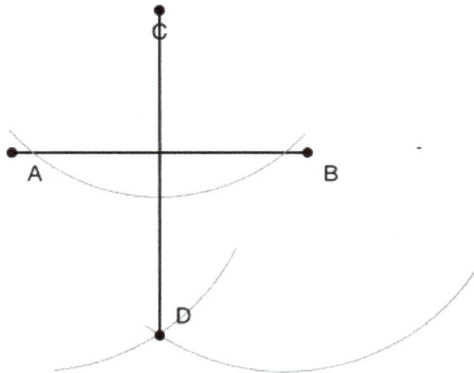

Types of symmetry

There are three types of symmetry:

- Line of symmetry which may also be called **reflective** symmetry or **mirror** symmetry is a line which cuts a shape in half so that it can be folded to cover the other half exactly.

- Translational symmetry - plane figures have translational symmetry if they can be translated (moved without turning) and still look the same.

- Rotational symmetry - plane figures have rotational symmetry of a certain order, if they fit onto themselves under rotations through given angles about a common centre.

- Order of symmetry – The order of symmetry indicates how many times an object fits back onto itself exactly as it is rotated through 360^0.

Lines of symmetry do not have to be horizontal or vertical. They can be in any direction. Rotational symmetry is not recognised in plane figures with rotational symmetry of order 1.

An object has point symmetry if it also has rotational symmetry of order 2. Point symmetry indicates that an object fits onto itself when rotated $180°$ about the same point.

> Plane figures may have more than one line of symmetry.

Example

How many lines symmetry does each shape have?

Equilateral triangle   Kite

> For regular polygons the number of lines of symmetry is equal to the number of the sides of the polygon.

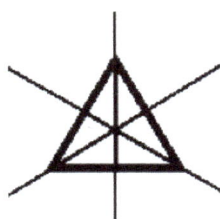

3 lines of symmetry   1 line of symmetry

Regular polygons are polygons that have all the interior angles equal and all the sides equal.

Solve geometric problems using properties

For straight lines, the equation of a straight line is in the form, $y = mx + c$ where m is the gradient of the line and c is the y-intercept. The gradient of the line is given by $m = \dfrac{y_2 - y_1}{x_2 - x_1}$

Example

L and M are two points on a straight line. L is the point $(3, 5)$ and M is the point $(6, 7)$.

(a) Determine the gradient.

$$m = \frac{y_2 - y_1}{x_2 - x_1} = \frac{7 - 5}{6 - 3} = \frac{2}{3}$$

(b) The intercept of LM on the y-axis.

The y-intercept of the line which passes through the points L(3, 5) and M(6,7) is found via

$$y = \frac{2}{3}x + c$$

Substituting $(3, 5)$ in the equation above gives

$$5 = \left(\frac{2}{3}\right)3 + c$$
$$5 = 2 + c$$
$$c = 3$$

Therefore the y intercept is 3.

(c) The equation of the line which passes through L and M.

The equation of the line that passes through the points L(3, 5) and M(6,7) is given by $y = \dfrac{2}{3}x + 3$

Properties of angles

Two angles are adjacent when they share a common vertex and a common arm (side) without overlapping. When a straight line transverses (crosses) one or more lines, the results are as follows:

- Opposite angles are equal.

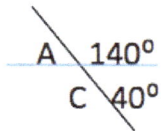

Angles A and C are adjacent.

Angle $A = 40°$ (vertically opposite angles)
Angle $C = 140°$. (vertically opposite angles)

- When a transversal crosses a pair of parallel lines, the alternate angles formed are equal; the corresponding angles are equal; and the co-interior angles (two angles in between the parallel lines that are on the same side of the transversal) sum to $180°$.

Examples

For the diagram shown below,

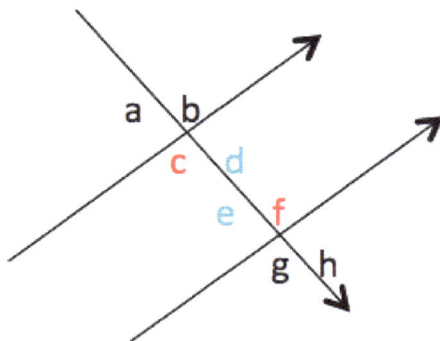

(a) Name two pairs of alternate angles.

c and f are alternate to each other.
d and e are alternate to each other.

(b) Name four pairs of corresponding angles.

a and e

b and f

c and g

d and h

(c) Name two pairs of adjacent angles.

c and d are adjacent to each other.

f and h are adjacent to each other.

(d) Name two pairs of co-interior angles.

d and f are co-interior angles. $\left(d+f=180°\right)$

c and e are co-interior angles. $\left(c+e=180°\right)$

Complementary angles

- Complementary angles sum to $90°$

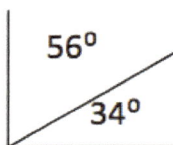

The two angles 56° and 34° are complementary.

- Supplementary angles sum to $180°$
-

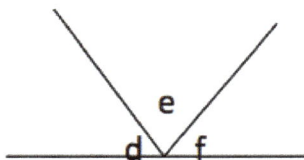

$d+e+f=180°$

The sum of the angles at a point sum to $360°$

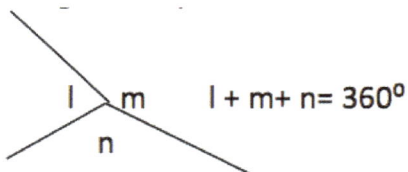

$l+m+n=360°$

Example

Determine the size of angle b.

$b = 360° - \left(60° + 70° + 50° + 50°\right)$

$b = 360° - 230°$

$b = 130°$

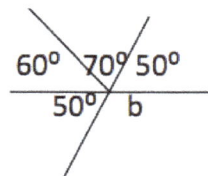

Properties of polygons

Polygons are closed plane figures having three or more sides.

- The sum of the interior angles of a regular n-sided polygon $= \left(n - 2\right) \times 180°$
- The sum of the exterior angles of a regular polygon is $360°$ or 4 right angles.

Examples

1.   Calculate the size of the interior angles of a regular pentagon.

Method 1

A pentagon has 5 sides

The sum of the interior angles of a pentagon $= \left(5 - 2\right) \times 180° = 540°$

So the size of one of the interior angle of a regular pentagon $= 540° \div 5 = 108°$

Method 2

A regular pentagon has 5 sides so the size of an exterior angle $= 360° \div 5 = 72°$

So for a regular pentagon, the size of the interior angles $= 180° - 72° = 108°$ (Angles on a straight line)

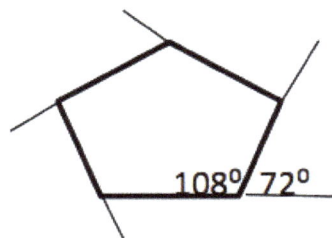

A regular polygon has all sides of equal length and all angles of equal size.

2.   How many sides does a regular polygon have if each interior angle is 157.5° ?

Method 1

Let the number of sides $= n$ for the regular polygon with interior angles of 157.5°

Recall: The sum of the interior angles $= (n - 2) \times 180°$

Thus $(n - 2) \times 180° = 157.5n$.

$180n - 360° = 157.5n$

$180n - 157.5n = 360°$

$22.5n = 360$

$n = \dfrac{360}{22.5} = 16$

Hence the regular polygon has 16 sides.

Method 2

The exterior angle of the regular polygon $= 180° - 157.5° = 22.5°$

The sum of the exterior angles of a polygon $= = 360°$

Hence the number of sides of this regular polygon $= \dfrac{360}{22.5} = 16.$

The regular polygon has 16 sides whose interior angle is 157.5°

Properties of circles

- A circle is a curved shape with each point on the line the exact distance from the centre.
- The circumference, C , is the perimeter of a circle.
- A radius, r , is a line drawn from the centre of the circle to the circumference.
- A diameter, d , is a straight line drawn from one end of the circumference passing through the middle of the circle to another point on the circumference. $D = 2r$.
- An arc is a part of the circumference of the circle.
- A chord is a straight line drawn from a point to another point on the circumference. The diameter of the circle is a special chord that divides the circle into two semicircles.

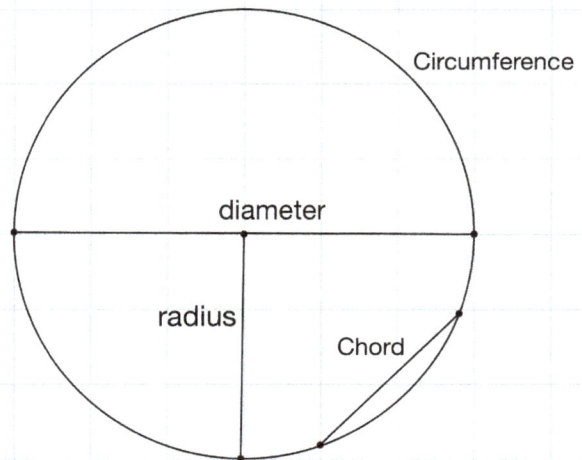

- A sector is an area of the circle that is bounded by an arc and two radii. In the figure below, the shaded area is the sector ABC.

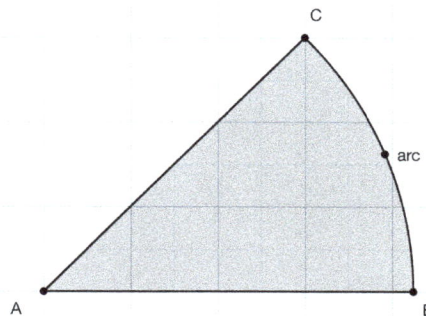

- A segment is a part of the circle that is bounded by a chord and the arc produced by the same chord.

- Equal Circles - Two circles are equal if they have the same radius.

Represent translations in the plane using vectors

The translation $T = \begin{pmatrix} x \\ y \end{pmatrix}$ means move x units horizontally and y units vertically.

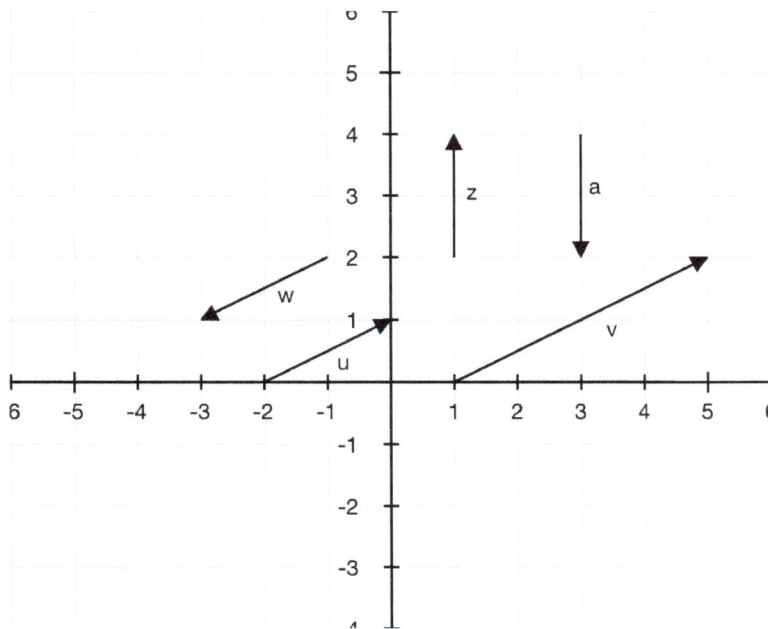

In the diagram above,

- The vector $a = \begin{pmatrix} 0 \\ -2 \end{pmatrix}$ represents a translation of 0 units to the right and 2 units down.

- The vector $z = \begin{pmatrix} 0 \\ 2 \end{pmatrix}$ and represents a translation of 0 units to the right and 2 units up.

- $a = -z$ and the vectors are said to be going in opposite direction to each other.

- The vector $w = \begin{pmatrix} -2 \\ -1 \end{pmatrix}$ represents a translation of 2 units to the left and 1 unit down.

- The vector $u = \begin{pmatrix} 2 \\ 1 \end{pmatrix}$ represents a translation of 2 units to the right and 1 up.

- $w = \begin{pmatrix} 2 \\ 1 \end{pmatrix} = -u$.

- The vector $v = \begin{pmatrix} 4 \\ 2 \end{pmatrix}$ represents a translation of 4 units to the right and 2 units upwards.

- $v = 2\begin{pmatrix} 2 \\ 1 \end{pmatrix} = 2u$

Determine and represent the location of an object or its image

Consider the following examples.

Examples

1.   The vertices of a triangle are $A\,(-3,\,3)$, $B\,(0,\,5)$ and $C\,(1,\,3)$. ΔABC is mapped onto $\Delta A'B'C'$ by a translation represented by the vector $T\begin{pmatrix} 5 \\ -2 \end{pmatrix}$. Determine the position of the vertices of the image triangle $\Delta A'B'C'$.

$A + T = A'$

$$\begin{pmatrix} -3 \\ 3 \end{pmatrix} + \begin{pmatrix} 5 \\ -2 \end{pmatrix} = \begin{pmatrix} -3+5 \\ 3-2 \end{pmatrix} = \begin{pmatrix} 2 \\ 1 \end{pmatrix}$$

$B + T = B'$

$$\begin{pmatrix} 0 \\ 5 \end{pmatrix} + \begin{pmatrix} 5 \\ -2 \end{pmatrix} = \begin{pmatrix} 0+5 \\ 5-2 \end{pmatrix} = \begin{pmatrix} 5 \\ 3 \end{pmatrix}.$$

$C + T = C'$

$$\begin{pmatrix} 1 \\ 3 \end{pmatrix} + \begin{pmatrix} 5 \\ -2 \end{pmatrix} = \begin{pmatrix} 1+5 \\ 3-2 \end{pmatrix} = \begin{pmatrix} 6 \\ 1 \end{pmatrix}$$

The vertices of triangle A' B' C' under the translation $\begin{pmatrix} 5 \\ -2 \end{pmatrix}$ are

A' (2, 1), B' (5, 3) and C' (6, 1).

The figure below shows the triangle ABC and its image $A'B'C'$ after the translation $\begin{pmatrix} 5 \\ -2 \end{pmatrix}$.

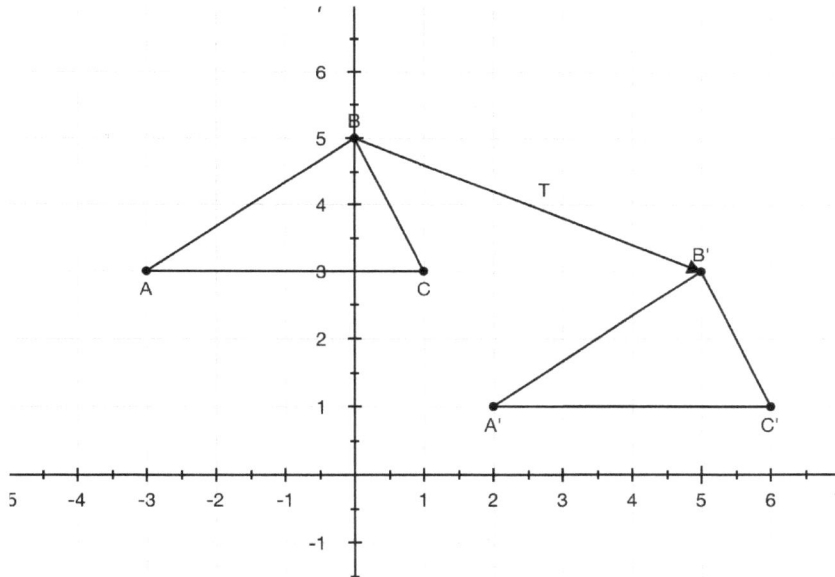

2.   Triangle ABC is mapped onto triangle $A'B'C'$ with vertices $A'(3,6)$ $B'(1,4)$ and C' (2, 0) under the translation $T = \begin{pmatrix} -2 \\ 6 \end{pmatrix}$. Determine the position of the vertices of triangle ABC.

$A' - T = A$

$$\begin{pmatrix} 3 \\ 6 \end{pmatrix} - \begin{pmatrix} -2 \\ 6 \end{pmatrix} = \begin{pmatrix} 3+2 \\ 6-6 \end{pmatrix} = \begin{pmatrix} 5 \\ 0 \end{pmatrix}$$

$B' - T = B$

$$\begin{pmatrix} 1 \\ 4 \end{pmatrix} - \begin{pmatrix} -2 \\ 6 \end{pmatrix} = \begin{pmatrix} 1+2 \\ 4-6 \end{pmatrix} = \begin{pmatrix} 3 \\ -2 \end{pmatrix}$$

$C' - T = C$

$$\begin{pmatrix} 2 \\ 0 \end{pmatrix} - \begin{pmatrix} -2 \\ 6 \end{pmatrix} = \begin{pmatrix} 2+2 \\ 0-6 \end{pmatrix} = \begin{pmatrix} 4 \\ -6 \end{pmatrix}$$

Thus the vertices of triangle ABC are $A\,(5, 0)$, $B\,(3, -2)$ and $C\,(4, -6)$.
The diagram below shows the translation T.

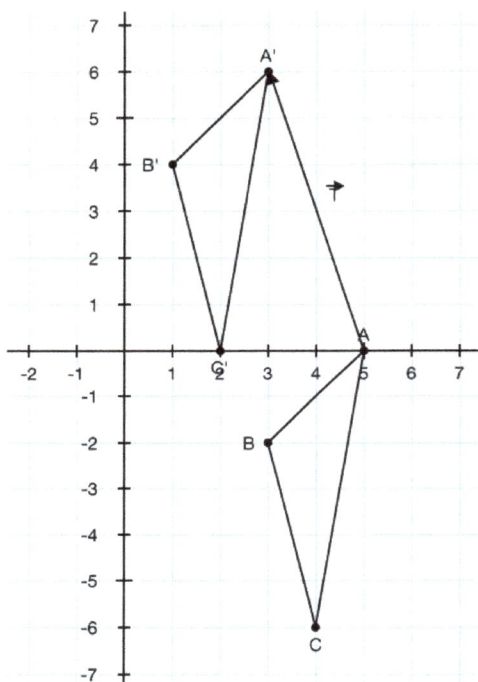

Geometric transformations

For a **rotation** through an angle about the origin, the
- lengths of the corresponding line segments are unchanged.
- corresponding angles are unchanged.
- object is congruent to its image.

For a **reflection** in a straight line or in the origin,
- the lengths of corresponding line segments are unchanged.
- corresponding angles are unchanged.
- the object is congruent to its image.

For a **translation**,
- the lengths of corresponding line segments are unchanged.
- corresponding angles are unchanged.
- the object is congruent to its image.

For **enlargements**, the
- centre of enlargement, a point and its image are collinear (on the same line).
- area of the image = $k^2 x$ the area of the object, where k is the scale factor of enlargement.
- ratio of the length of the line segment from the centre of enlargement to the image point and the length of the line segment from the centre of enlargement to corresponding object point is a constant called the scale factor, k.

Example

The diagram below shows parallelogram $ABCD$ under a rotation of $90°$ anti clockwise about the origin. State the relationship between $ABCD$ and its image $A'B'C'D'$.

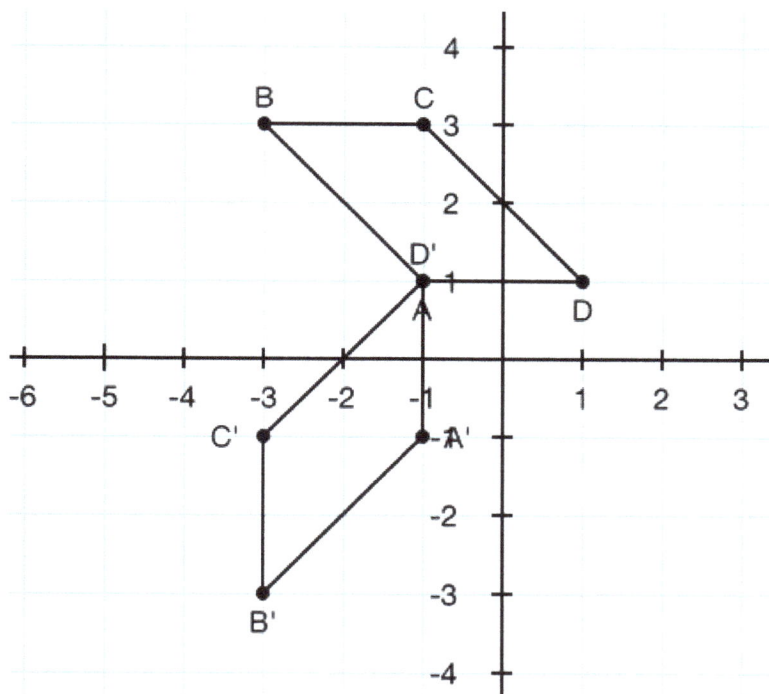

Parallelogram $ABCD$ is congruent to its image. The lengths of the corresponding line segments are unchanged. The corresponding angles are unchanged.

Describe a transformation given an object and its image.

- To describe any **enlargement** state the centre of enlargement and the scale factor of enlargement. The scale factor, k, determines if the enlargement is a reduction, 'reversed' reduction, 'reversed' enlargement or enlargement.
- For a reduction, the scale factor is a value between 0 and 1.
- For a 'reversed' reduction, the scale factor is a value between -1 and 0.
- For an enlargement, the scale factor is a value greater than 1.
- For a 'reversed' enlargement, the scale factor is a value less than -1.
- The term 'reversed' enlargement means the object and the image are on different sides of the centre of enlargement and the image is bigger than the object.
- To describe a **rotation** state the centre of rotation, the direction of rotation and the angle through which the object is rotated.
- To describe a **translation** state the displacement of the object in line with the x-axis and the y-axis. This can be written in the form of a column vector such as $\begin{pmatrix} c \\ d \end{pmatrix}$ where c represents the displacement parallel to the x-axis and d, the displacement parallel to the y-axis.
- To describe a **reflection** state the axis of reflection or the mirror line of reflection.
- To describe a glide reflection state the axis of reflection or the mirror line of reflection and the translation.

Examples

1. The graph below shows triangle ABC and its image $A'B'C'$ after an enlargement.

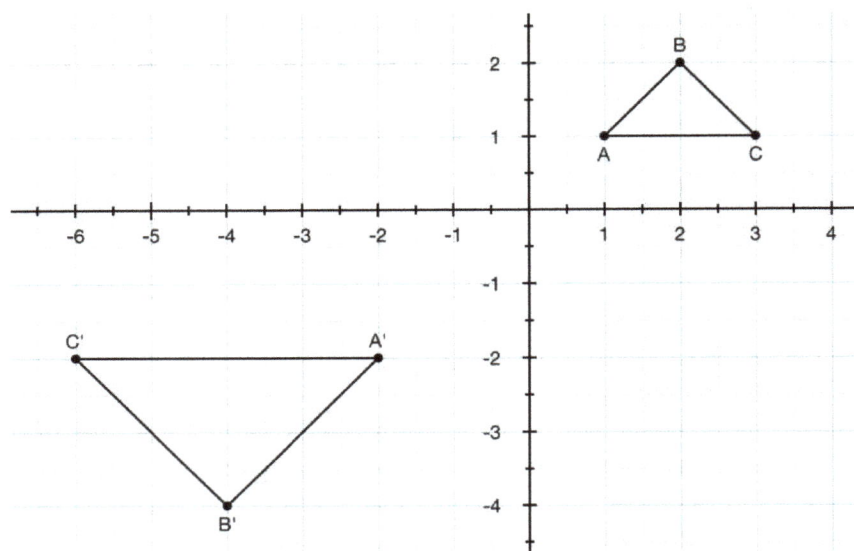

(a) Describe the transformation fully.

Connect the point A and its image A'; connect the point C and its image C'. The point of intersection of these two lines gives the centre of enlargement. (See the graph below). The coordinates of the centre of enlargement is the origin $(0, 0)$.

‣ Length of $OA = \sqrt{2}$
‣ Length of $OA' = 2\sqrt{2}$ (Pythagoras theorem)
‣ Since OA' is in the opposite direction to OA we write OA' as $2\sqrt{2}$
‣ Scale Factor, $k = \dfrac{OA'}{OA} = \dfrac{-2\sqrt{2}}{\sqrt{2}} = -2$

The transformation is a 'reversed' enlargement of scale factor -2 with centre of enlargement $(0, 0)$.

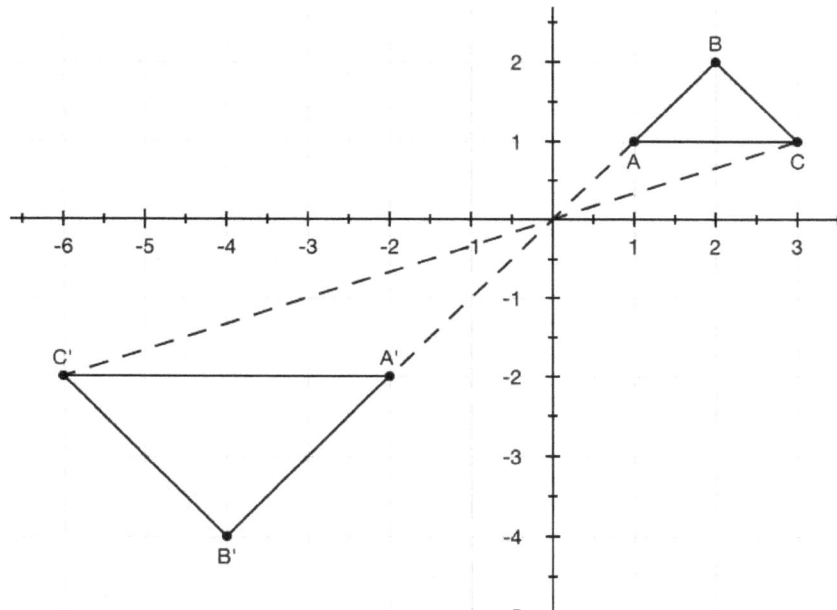

(b) Calculate the area of ABC and the area of $A'B'C'$.

Area of ABC $= \dfrac{1}{2} \times 2 \times 1 = 1$ unit²

Area of A'B'C' $= \dfrac{1}{2} \times 4 \times 2 = 4$ unit²

(c) Determine the ratio of the area of ABC to the area of $A'B'C'$.

The ratio of the area of ABC to the area of $A'B'C' = 1:4$. The image is four times as big as the object.

2.   The diagram below shows a square $ABCD$ and its image $A'B'C'D'$.

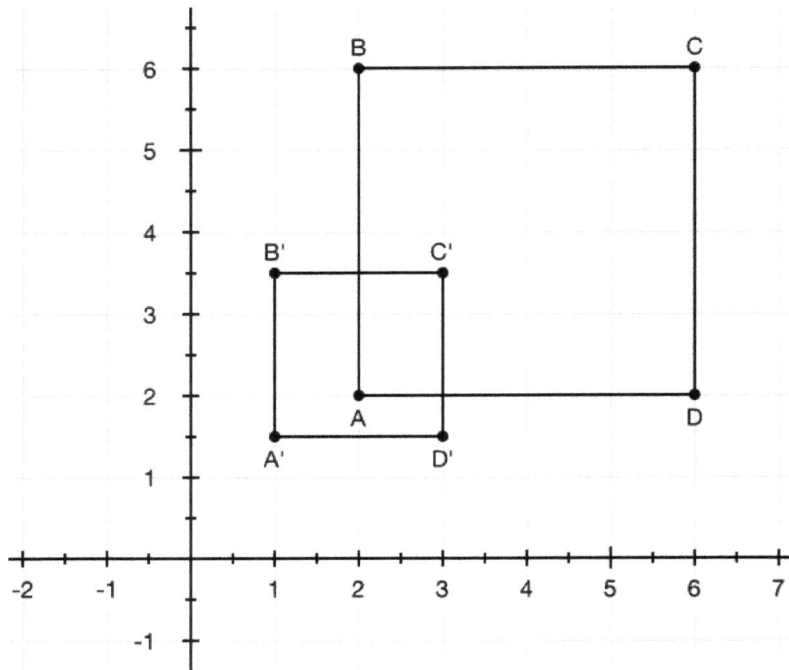

(a) Find the centre of enlargement clearly showing the method.

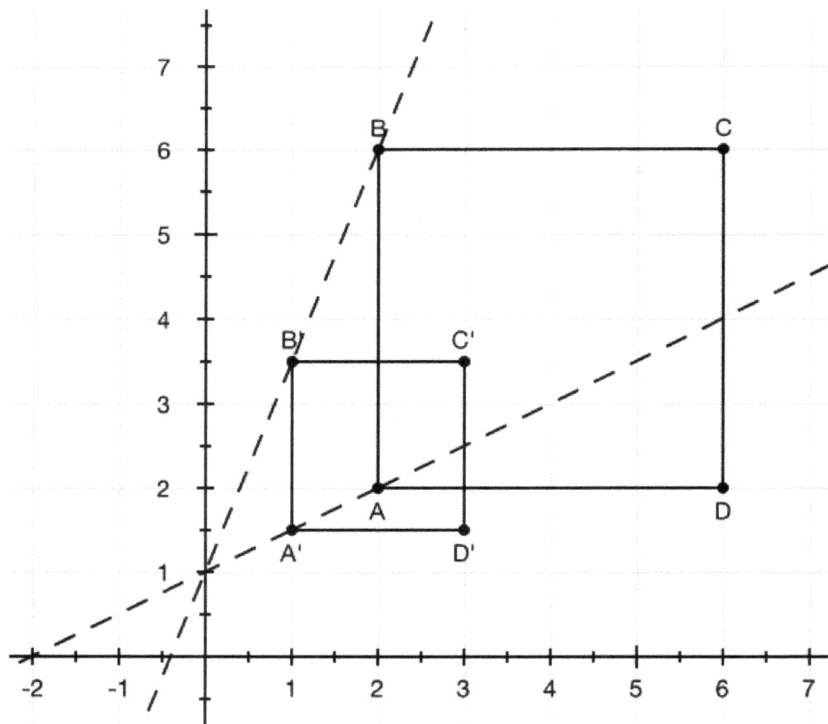

(b) Write the coordinates of the centre of enlargement

From the diagram, the centre of enlargement is $(0, 1)$.

(c) State the scale factor of enlargement.

Scale Factor, $k = \dfrac{PB'}{PB} = \dfrac{PA'}{PA} = \dfrac{PC'}{PC} = \dfrac{PD'}{PD} = \dfrac{1}{2}$.

(d) Find the ratio of the area of $A'B'C'D'$ to the area of $ABCD$.

Area of $A'B'C'D' = 2\ units \times 2\ units = 4\ units^2$.
Area of $ABCD = 4\ units \times 4\ units = 16\ units^2$.
The ratio of the area of $A'B'C'D'$ to $ABCD = 4:16 = 1:4$.

3.   In the diagram below triangle ABC and its image $A'B'C'$ are shown. Describe the transformation that maps ABC unto $A'B'C'$.

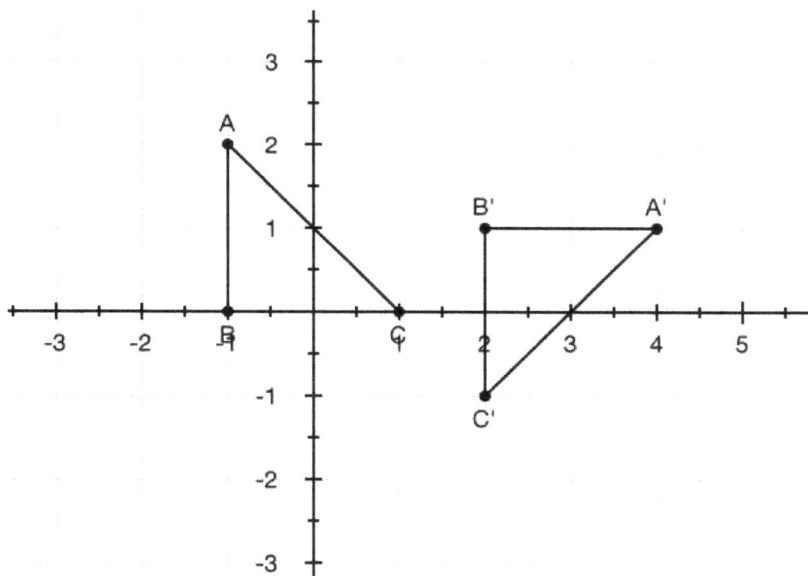

The centre of rotation is $D\ (1, -1)$. Triangle ABC is rotated $90°$ clockwise onto its image.

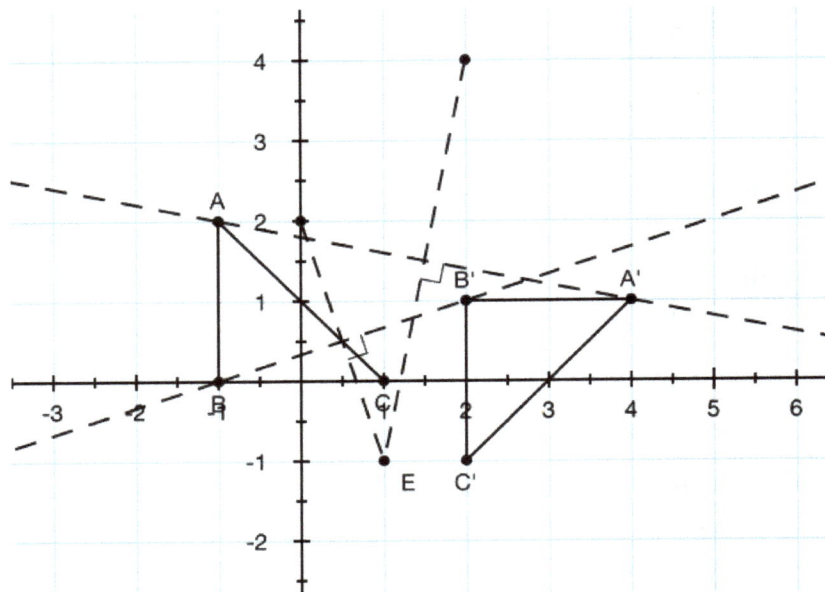

Locate the image of a set of points under a combination of transformations

Transformations may alter the position, shape or orientation of an object. When performing more than one transformation the order in which the transformations are performed is important.

Example

Plot the points $A\,(0, 3), B(3, 3), C(3, 1)$ and draw the triangle. Translate the triangle by the vector $\begin{pmatrix} -2 \\ -3 \end{pmatrix}$ and reflect the image $A'B'C'$ in the line $x = 2$.

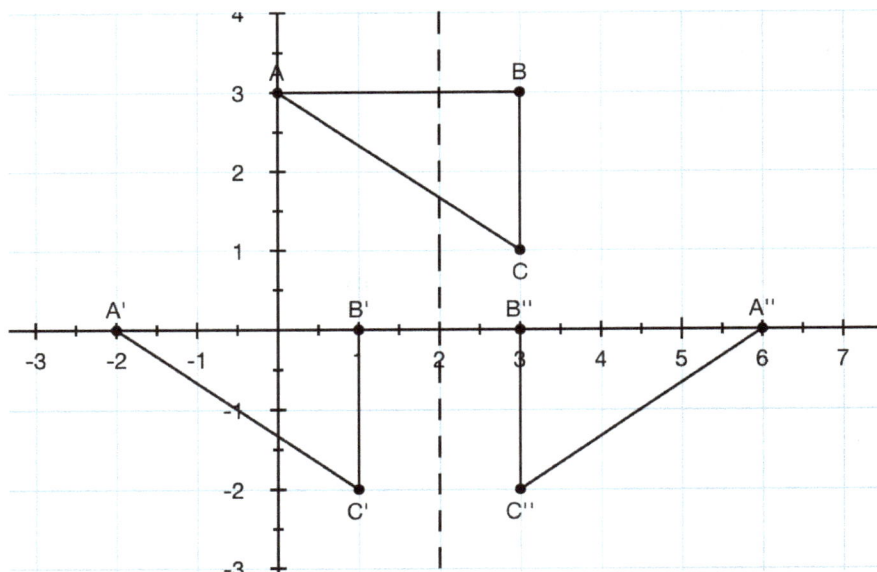

The triangle $A'B'C'$ is the image of ABC after the translation $\begin{pmatrix} -2 \\ -3 \end{pmatrix}$ and the triangle $A''B''C''$ is the image of the reflection of $A'B'C'$ in the line $x = 2$.

State the relations between an object and its image

Example

State two relationships between triangle ABC and its image $A''B''C''$ in the figure below.

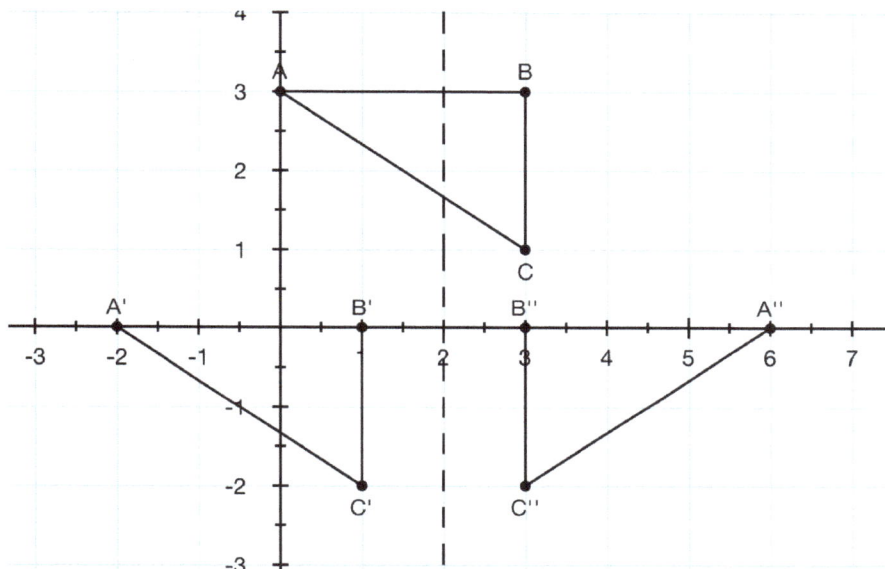

In the diagram above, ABC underwent a translation followed by a reflection. In doing so the corresponding angles of ABC and $A''B''C''$ were invariant and ABC is congruent to $A''B''C''$.

Use Pythagoras' theorem to solve problems

Pythagoras' theorem is used to determine an unknown side of a right triangle when two sides are given. Pythagoras' Theorem states the square of the hypotenuse is equal to the sum of squares of the two other sides: $a^2 + b^2 = c^2$, where a, b and c are the lengths of the sides of a right triangle and c is the hypotenuse.

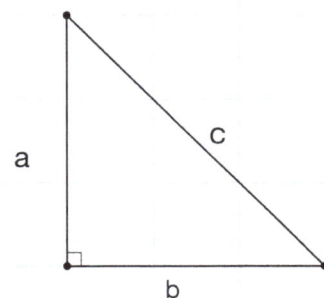

Example

Sketch the following triangles and determine the length of the unknown side.

(a) ΔDEF, $DE = 10\ cm$, $EF = 4cm$, $\hat{E} = 90°$. Find DF.

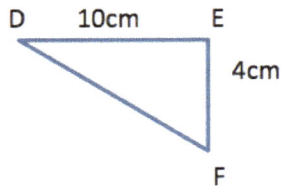

$DF^2 = DE^2 + EF^2$ (Pythagoras' theorem)
$\qquad = 10^2 + 4^2$
$\qquad = 116$
$DF = \sqrt{116} = 10.77\,\text{cm}$

(b) ΔDEF, $DF = 17\ cm$, $EF = 15cm$, $\hat{E} = 90°$. Find DE.

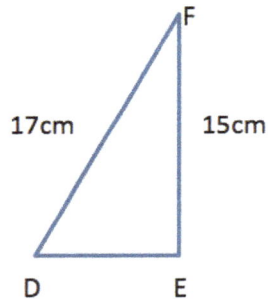

$DF^2 = DE^2 + EF^2$ (Pythagoras' theorem)
$EF^2 = DF^2 - EF^2$
$DE^2 = 17^2 - 15^2$
$\quad = 64$
$DE = \sqrt{64} = 8\text{cm}$

(c) ΔDEF, $DF = 12\ m$, $DE = 6cm$, $\hat{E} = 90°$. Find EF.

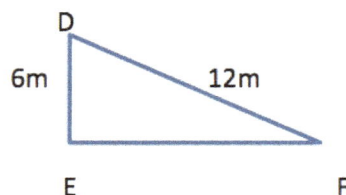

$DF^2 = DE^2 + EF^2$ (Pythagoras' theorem)

$EF^2 = DF^2 - DE^2$

$EF^2 = 12^2 - 6^2$

$= 108$

$EF = \sqrt{108} = 10.39\text{cm}$

Determine the trigonometric ratios of acute angles in a right-angled triangle

There are three trigonometric ratios that are used to calculate unknown sides or lengths in right-angled triangles. For the angle θ under consideration

- $\sin\theta$ = ratio of the side opposite the angle θ to the hypotenuse.
- $\cos\theta$ = ratio of the side adjacent to the angle θ to the hypotenuse.
- $\tan\theta$ = ratio of the side opposite the angle θ to the side adjacent to angle θ.

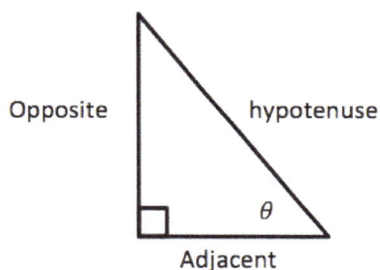

Example

Based on the figure below, determine $\sin\theta$, $\cos\theta$, $\tan\theta$

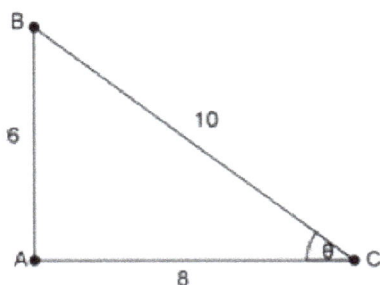

$$\sin\theta = \frac{6}{10} \qquad \cos\theta = \frac{8}{10} \qquad \tan\theta = \frac{6}{8}$$

Use trigonometric ratios in the solution of right-angled triangles

In a right angle triangle, the hypotenuse is always the longest side; the opposite side is the side opposite the angle being used and the adjacent is the third side of the triangle that was not mentioned.

Example

1. Calculate the distance AB in the figure below not drawn to scale.

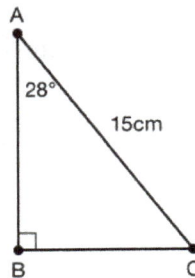

$$\cos 28° = \frac{AB}{AC} = \frac{AB}{15}$$

Thus $AB = 15 \cos 28°$

$\qquad = 15 \times 0.883$

$\qquad\quad = 13.25cm$ (Correct to 2 d.p.)

2. Calculate the distance AB in the figure shown below.

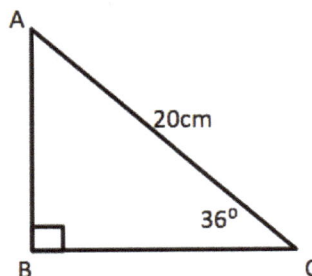

$$\sin 36° = \frac{AB}{AC} = \frac{AB}{20}$$

Thus $AB = 20 \sin 36°$

$= 20 \times 0.588$

$= 11.76cm$ (Correct to 2 d.p.)

3.   Calculate the distance AB in the figure shown below.

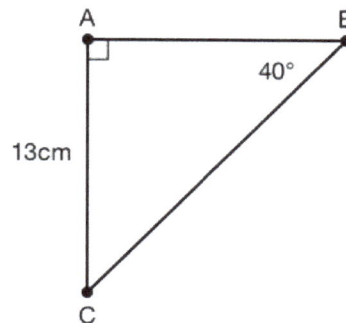

$$\tan 40° = \frac{AC}{AB} = \frac{13}{AB}$$

$$AB \times \tan 40° = 13$$

$$AB = \frac{13}{\tan 40°} = \frac{13}{0.839}$$

$$= 15.49 cm \qquad \text{(correct to 2 d.p)}$$

Use trigonometric ratios to solve problems based on physical world measures

- The angle of elevation is the angle between the horizontal and our line of sight as we look up to view an object.

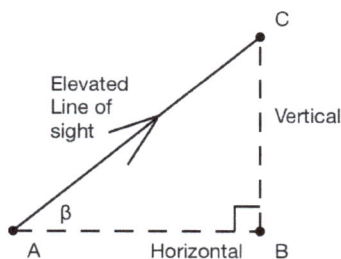

- The angle of depression is the angle between the horizontal and our line of sight, as we look down to view an object.

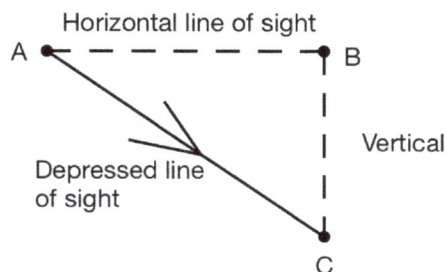

Examples

1.   The angle of elevation of the top of a flagpole from a point $50m$ from the base of the flagpole is $56°$. How high is the flagpole?

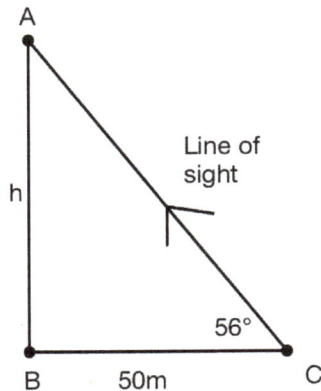

> The angle of depression is an alternate angle to the angle of elevation .

$$\tan 56° = \frac{h}{50}$$

$$h = 50 \tan 56°$$

$$= 74.13$$

Thus, the height of the flagpole is $74.1m$ correct to 1 d.p.

2.   From the top of a cliff $60m$ high, the angle of depression of a ship is $42°$. How far from the base of the cliff is the ship?

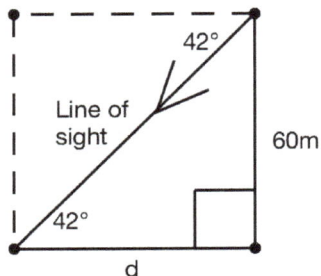

$$\tan 42° = \frac{60m}{d}$$

$$d = \frac{60m}{\tan 42^0}$$

$$= 66.64m$$

The ship is $66.64m$ from the base of the cliff.

• sine rule $$\frac{a}{\sin A} = \frac{b}{\sin B} = \frac{c}{\sin C}$$

• cosine rule $$a^2 = b^2 + c^2 - 2bc\ \cos A$$

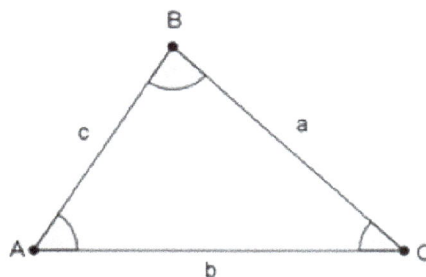

The **sine rule** is used for triangles when two angles and a side opposite one of the angles is known and you are required to find the length of the side opposite the other known angle. It can also be used in cases where two sides and one the angles opposite a known angle is given and you are required to find the angle opposite the other known side.

The **cosine rule** is used to calculate a side of a triangle when the two other sides and the angle opposite the side to be determined are given. It may also be used to calculate an angle in a triangle given the length of all three sides of the triangle.

Examples

1.   In the diagram below, not drawn to scale, $AC = 9cm$, angle $ABC = 50°$, angle $BAC = 41°$. Calculate the length of BC giving your answers to 1 d.p.

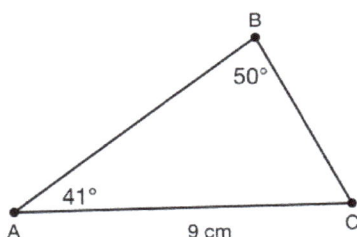

$AC = b$, $BC = a$, $B = 50°$, $A = 41°$.

Using the sine rule $\dfrac{a}{\sin A} = \dfrac{b}{\sin B}$

$$\frac{a}{\sin 41°} = \frac{9}{\sin 50°}$$

$a = \dfrac{\sin 41 \times 9}{\sin 50°} = 7.71cm$. Therefore $BC = 7.7cm$ correct to 1 d.p.

2.   In triangle *DEF,* not drawn to scale, $DE = 20cm$, $EF = 14cm$ and $< E = 36°$.

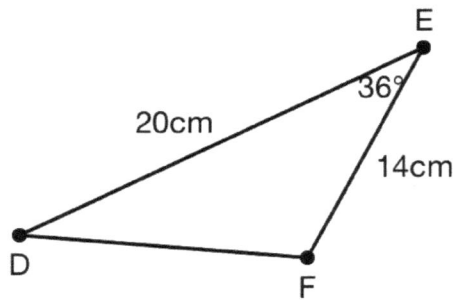

(a) Calculate the length of DF.

Applying the cosine rule

$DF^2 = 20^2 + 14^2 - 2\,(20 \times 14)\,\cos 36°$

$= 400 + 196 - 560\,\cos 36°$

$= 142.95$

$DF = \sqrt{142.95} = 11.96$

(b) Use the cosine rule to find the size of $< F$.

Applying the cosine rule

$\cos F = \dfrac{11.96^2 + 14^2 - 20^2}{2\ x\ 11.96\ x\ 14}$

$= -0.182$

$F = \cos^{-1}(-0.182) = 100°$

correct to the nearest degree

Represent the relative position of two points given the bearing of one point

A bearing is the position of an object relative to the position of another object. The bearing is obtained by measuring the angle in a clockwise direction from the north line towards the line joining both objects.

Examples

1.   Represent the following information on separate sketches.

(a) The bearing of B from A is 135°.

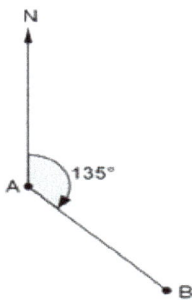

Bearings should always be written with three figures.

(b) The bearing of D from C is 065°.

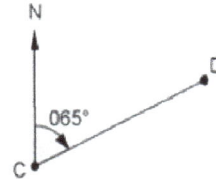

(c) The bearing of F from B is 281°.

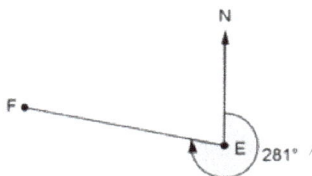

2.   The bearing of A from B is 135°. What is the bearing of B from A?

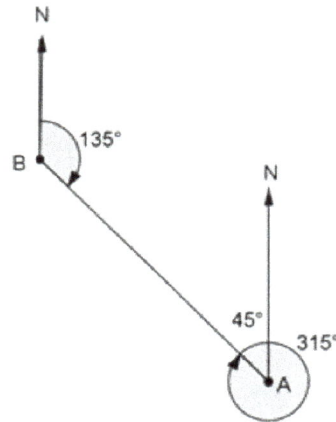

Since *NB* and *NA* are parallel then *NBA* and *NAB* are supplementary (co-interior),
So $< NAB = 180° - 135° = 45°$.
The bearing of *B* from *A* is $360° - 45° = 315°$. (<s at a point)
Hence the bearing of *B* from *A* is $315°$.

Solve geometric problems using properties of circles and circle theorems
Illustrated via the following examples.

Examples

1.   The angle, which an arc of a circle subtends at the centre of a circle, is twice the angle it subtends at any point on the remaining part of the circumference. Calculate the size of angle *XYZ* in the diagram below.

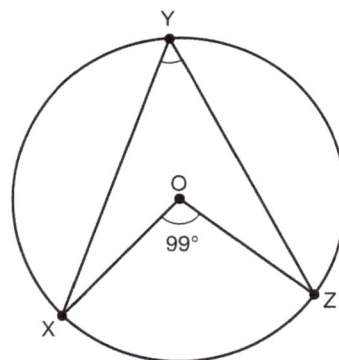

$< XYZ = \dfrac{1}{2} < XOZ$ (< at the centre of circle = 2 x angle subtended at circumference on same arc)

$= \dfrac{1}{2} \times 99° = 49.5°$

2.   In the diagram below, determine the magnitude of Angle ACB and Angle AEB giving reasons in each case.

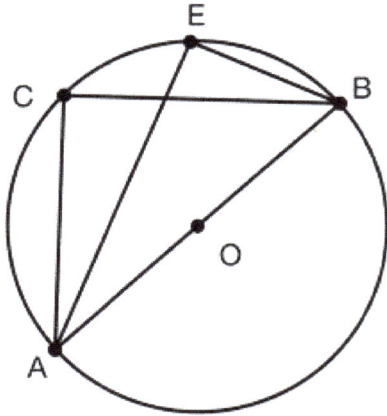

The angle in a semicircle is a right angle.

Now $< ACB = 90°$ ($<$ in a semi-circle)
Hence the magnitude of angle ACB is $90°$
Similarly $< AEB = 90°$ ($<$ in a semi-circle).
Hence the magnitude of angle AEB is $90°$

3.   In the figure below, Angle $< CAB = 50.91°$ and angle $ABD = 29.51°$. Find the size of the Angle ACD and Angle CDB

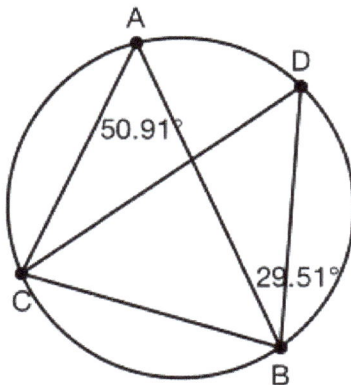

Angles in the same segment of a circle subtended by the same arc are equal.

The opposite angles of a cyclic quadrilateral are supplementary.

Angle ACD = Angle $ABD = 29.51°^0$ ($<$ in same segment subt. by same arc)
Angle CDB = Angle $CAB = 50.91°$ ($<$ in same segment subt. by same arc)

Questions
Geometry & Trigonometry

Lawrence Bishop

The St. Michael School Barbados

Jerome Stuart

Graydon Sealy School Barbados

QUESTIONS

[1] Calculate the length of the side w.

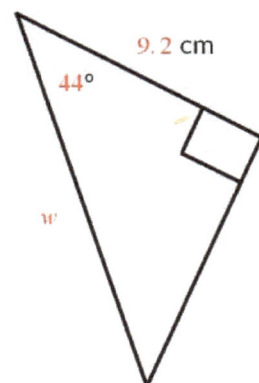

9.2 cm

44°

w

[2] A person 100 meters from the base of a tree observes that the angle between the ground and the top of the tree is 18 degrees. Estimate the height h of the tree to the nearest tenth of a meter.

[3] Calculate the size of the angle c.

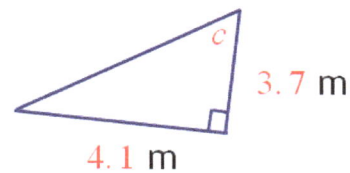

3.7 m

4.1 m

[4] In the diagram below, not drawn to scale, calculate to one decimal place
(a) The length of QT.
(b) The length of RT.
(c) The angle RST

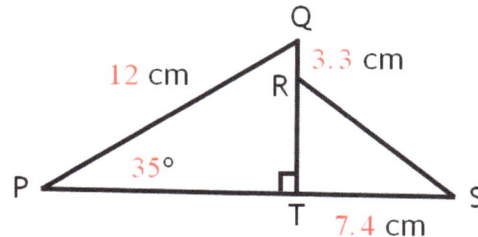

Q

12 cm   3.3 cm
                  R

35°

P

T   7.4 cm   S

[5] Calculate the size of the angle b.

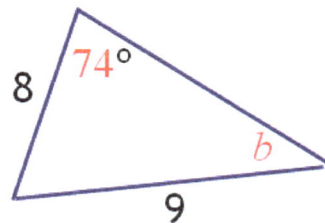

74°

8

b

9

[6] A ladder 18 ft long, leans against a vertical wall. Its foot is 8 ft from the foot of the wall. Calculate the angle between the ladder and the wall.

[7] Find the length of the side a.

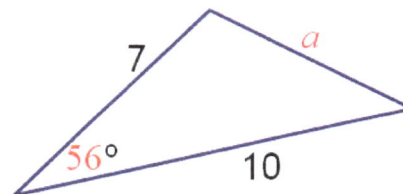

7   a

56°   10

[8] A man 1.2 metres tall stands 36 metres away from the foot of a tree. He observes the angle of elevation of the top to be 23 degrees from his line of sight. How tall is the tree?

[9] An airplane is flying at a height of 7 miles above the ground. The distance along the ground from the airplane to the airport is 12 miles. What is the angle of depression from the airplane to the airport?

[10] John wants to measure the height of a building. He walks exactly 40 m from the building and looks up. The angle from the ground to the top of the building is 30 degrees. How tall is the building?

[11] A tree 70ft in height cast a shadow of length 60ft. what is the angle of elevation from, the end of the shadow to the top of the tree?

[12] Identify the opposite side, adjacent side and hypotenuse of the right-angled triangles shown below.

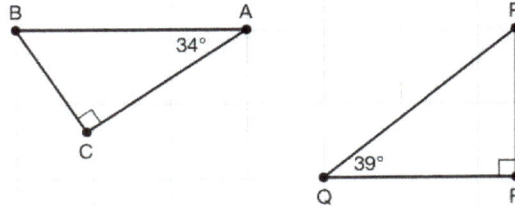

[13] A figure not drawn to scale represents a ship sailing from a harbor H on a bearing of $280°$ for 4km to A and then on a bearing of $060°$ for 12km to arrive at Z.

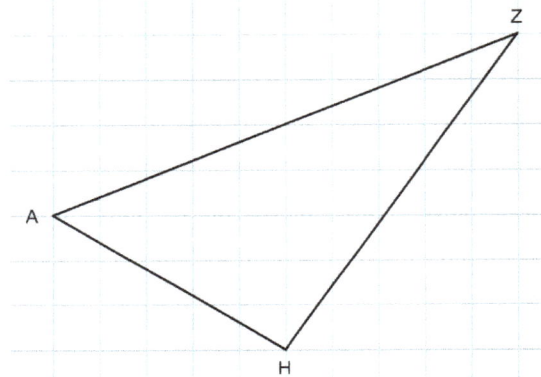

(a) Copy and complete the diagram showing clearly the bearings $280°$ and $060°$.
(b) Calculate the size of the angle HAZ.
(c) Calculate to the nearest whole number, the distance HZ.

[**14**] Determine angle *CEO* .

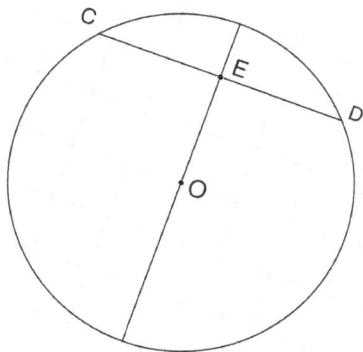

The line joining the centre of a circle to the midpoint of a chord is perpendicular to the chord.

[**15**] Determine the size of the angle *BAD* in the diagram above.

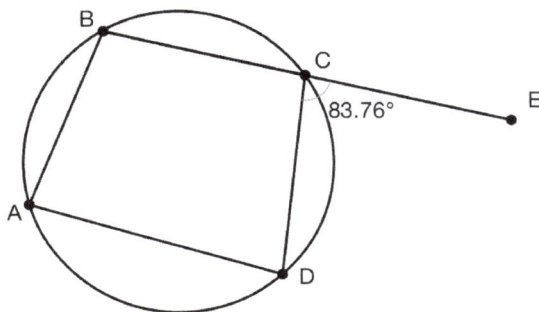

The exterior angle of a cyclic quadrilateral is equal to the interior opposite angle.

[**16**] The bearing of *D* from *C* is $300°$. What is the bearing of *C* from *D*?

[**17**] Calculate the bearing of port L from port P if a ship leaves port P and travelled 50 km due East and 35 km South to arrive at port *L*.

[**18**] Points *B*,*C* and *D* are points on the circumference of the circle, with centre, *O*. Angle $< BAC = 78°$. *AB* and *AC* are tangents to the circle at the points *A* and *C* respectively. Determine the size of angle *CBA*, *CBO*, *COB*, *CDB*

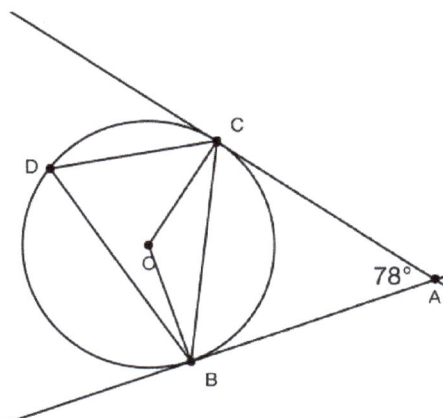

A tangent of a circle is perpendicular to the radius of that circle at the point of contact.

The lengths of two tangents from an external point to the points of contact on the circle are equal.

[19] In the figure below, not drawn to scale, there is a tangent at A and angle $CAB = 33°$. Determine the size of angle ADB, angle DAB, and angle ABD giving reasons for your answers.

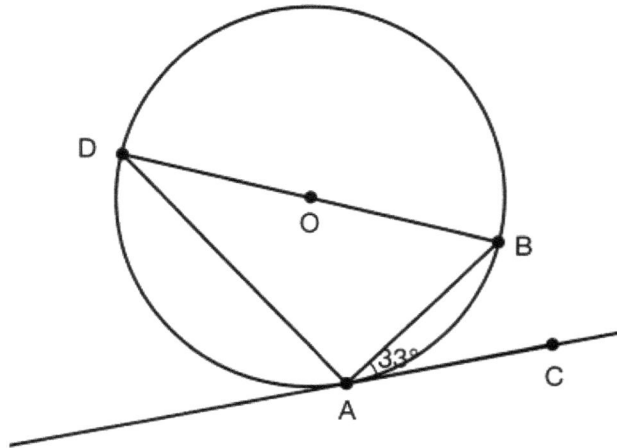

> The angle between a tangent to a circle and a chord through the point of contact is equal to the angle in the alternate segment.

[20] In the figure below, angle $DAB = 89.1°$ and angle $ABC = 104.69°$. Calculate, given a reason for each result, Angle CDA and Angle BCD.

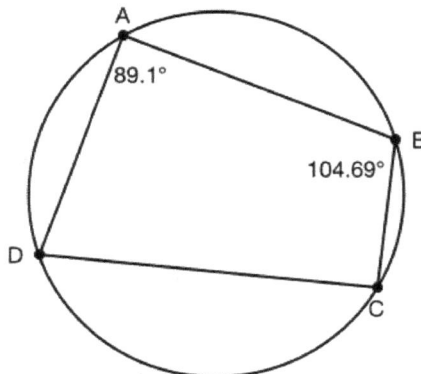

Chapter 8
Relations, Functions and Graphs
Charles VanderPool & Kemar Trotman
The Lodge School
Barbados

Events can be modelled by a function and represented on a graph. Functions are a very important aspect in mathematics, because they can be used to help describe and predict the outcome of events. Understanding this concept will help you in other subject ares such as Economics, Physics and Chemistry, just to name a few.

We hope this chapter will help to improve of your understanding of relations, functions and graphs.

Definitions of relations, functions and graphs

A relation is a rule that links two numbers together. Types of relations are,

- **One to one**
   When one number is linked to only one number.
   If 5 can only be linked to 8.

- **One to many**
   When one number is linked to more than one number.
   If 5 can be linked to other numbers besides 8.

- **Many to one**
   When many numbers is linked to only one number.
   If 5 along with some other numbers are linked to only 8.

Key terms

These can be expressed as x.

- **The domain**
   The domain is the set of possible numbers that are to be linked to another number.

- **Image**
   The image is the value that a number from the domain is linked to.

- **The range**
   The set containing the image points.

- **Co-domain**
   The set of possible numbers that the domain can be linked to is called the co-domain.

These can be expressed as f(x) or y

A relation can be represented in many ways, for example,

- **A set of ordered pairs**

    A set of ordered pairs consist of a domain element and its image.
    {(2, 5) (3,6) (4,7) (5,8)}

- **Arrow diagrams**

    In an arrow diagram, the domain points are grouped on the left and the image points are grouped on the right. An arrow is then drawn from each domain element to its image.

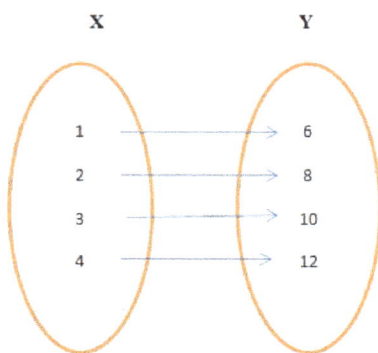

> Each group is encircled by an oval and written vertically, no number should be repeated and written in ascending order.

- **Graphically**

    Plot each ordered pair as co-ordinates. The graph below is a plot of the line $y = 2x + 3$

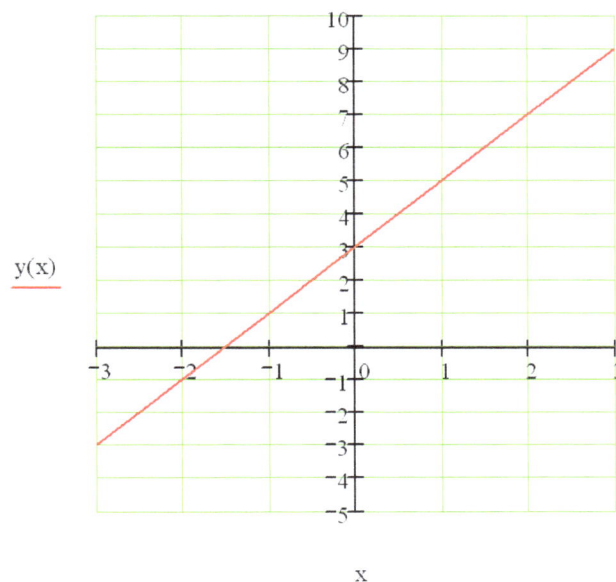

- **Algebraically**

$x \rightarrow 2x + 3$

> x are the domain values and each answer from 2x+3 would be the image values.

Characteristics that define a function

A function is a relation where each domain value is linked to one and only one value in the range.

Functional notation

Functional notational are ways of expressing a function. For example,

$f : x \rightarrow x^2$ or $f(x) = x^2$ as well as $y = f(x)$

> Functions are one to one or many to one relations

Example

Determine the range of $f(x) = 2x + 4$ for the domain $\{1, 2, 3, 4\}$

$f(1) = 2(1) + 4 = 2 + 4 = 6$
$f(2) = 2(2) + 4 = 4 + 4 = 8$
$f(3) = 2(3) + 4 = 6 + 4 = 10$
$f(4) = 2(4) + 4 = 8 + 4 = 12$

Range = {6, 8, 10, 12}

> For x = 1, 2x means 2 multiply by 1 and **not** 21.

> To use functional notation, replace x in the function, with each domain value.

Difference between a function and a relation

- **Arrow diagrams**

    If each member on the left is connected to only one member on the right then it can be determined as a function. However, if one member on the left is connected to more than one member on the right then it is **not** a function but a relation. For example,

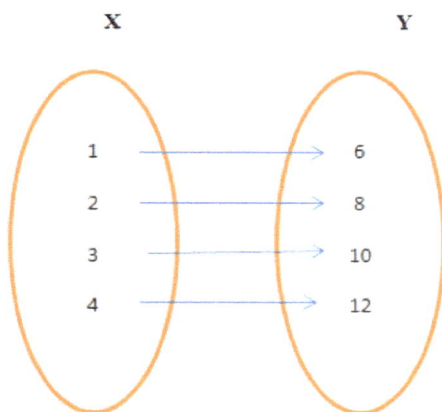

One to one

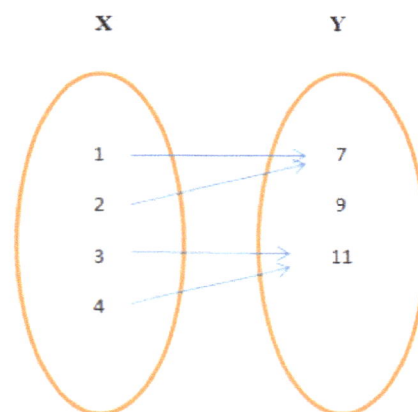

Many to one

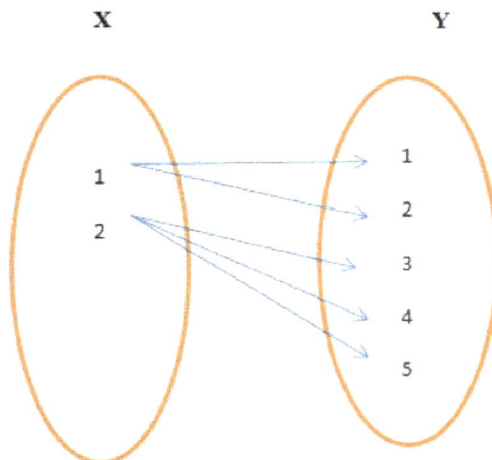

Notice one number is linked many numbers, therefore, this is **not** a function.

- **Ordered pairs**

    When a function is written as a set of ordered pairs, the values on the **left** of the pair should only be written once as left side values of the pair.

    ▸ Example of a function $\{(1,2)\ (2,3)\ (3,4)\ (4,5)\}$
    ▸ Example of a non-function $\{(1,2)\ (1,3)\ (2,4)\ (2,5)\}$

- **Graphically**
  If a graph of a function can pass the vertical line test then it is a function.

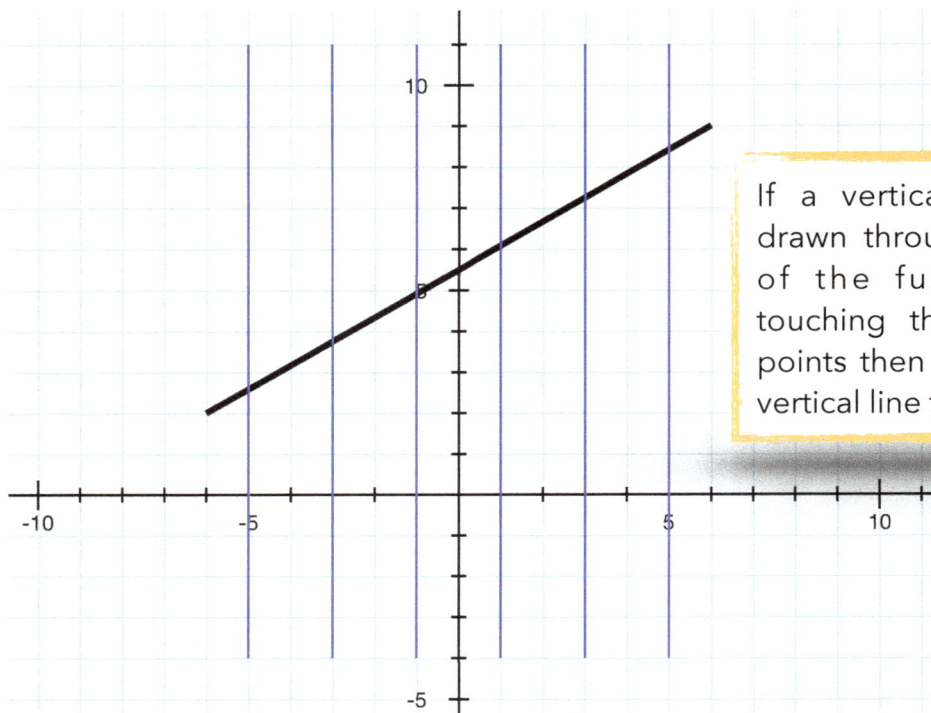

If a vertically line can be drawn throughout the graph of the function without touching the graph at two points then it has passed the vertical line test.

Each blue line only touches the graph once.

Linear function

A linear function is any function that can be drawn as a straight line. There are basically three ways to represent a linear function.

- $y = c$ where c is a real number. For example, $y = 4$

  These are always horizontal lines.

- $x = k$ where k is a real number. For example, $x = 7$

  These are always vertical lines.

- $y = mx + c$ where m and c are real numbers.
  For example, $y = 3x + 2$

  These are always diagonal lines.

Example

1.   Draw the line $y = 4$

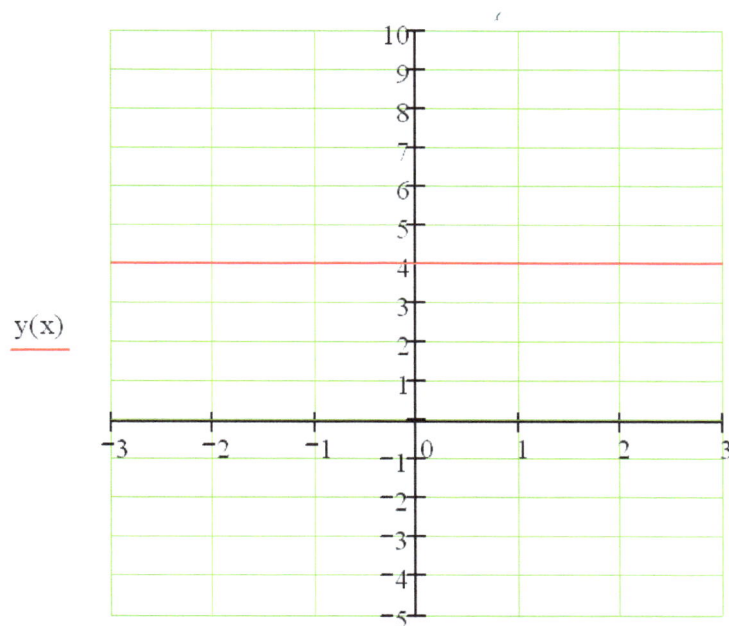

2.   Draw the line $x = 7$

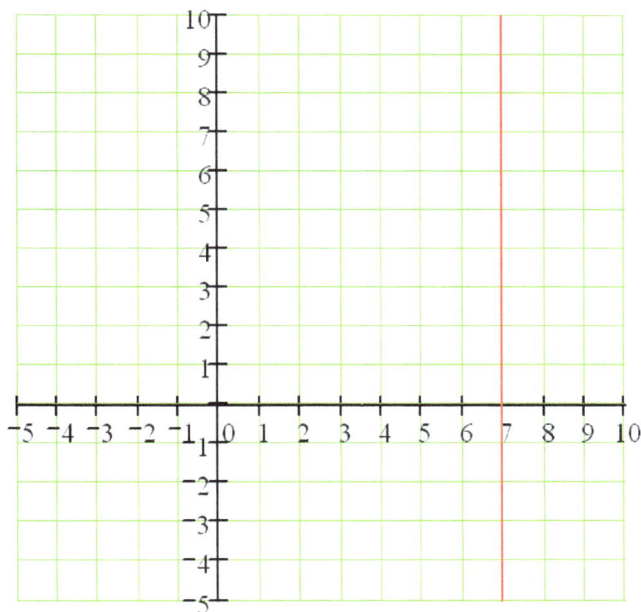

3.   Draw the line $y = 2x + 1$

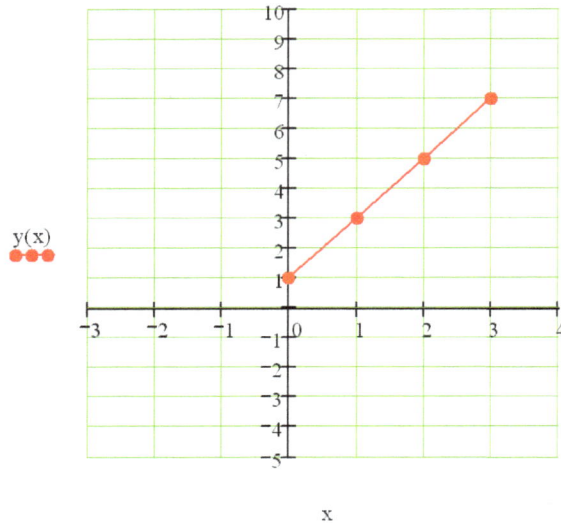

coordinates are
(0,1) (1,3) (2,5) (3,7)

Ordered pairs

Each graph can be read to reveal different ordered pairs. These ordered pairs are the coordinates of the different points on the graph. They show which range value is associated with which domain value.

Example

Given the domain value 2, what is its range value in the graph shown below.

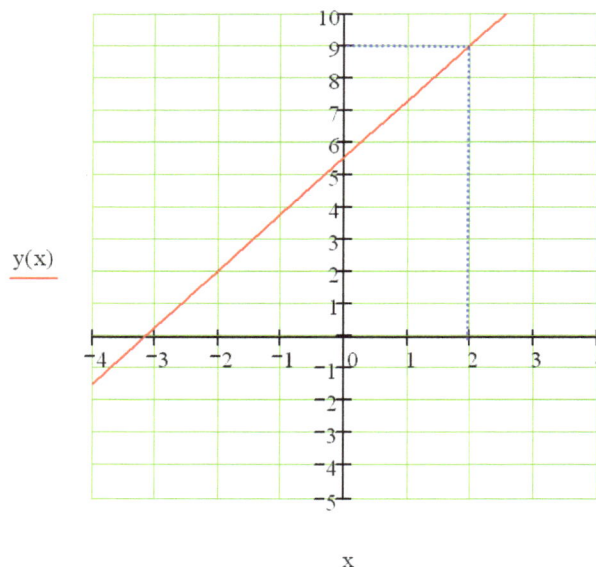

The range value associated with the domain value of 2 is 9. Therefore, one of the ordered pairs is (2,9).

Straight line equation

The equation of a straight line is given by $y = mx + c$, where m is the gradient and c is the intercept on the y-axis. The intercept c is easily found by setting $x = 0$, because $y = m(0) + c = c$.

Example

Find the intercept of the line $y = 2x + 3$.

Set $x = 0$ to find the intercept $y = 2(0) + 3 = 3$
Hence $c = 3$. A plot of this line is shown below to confirm the answer.

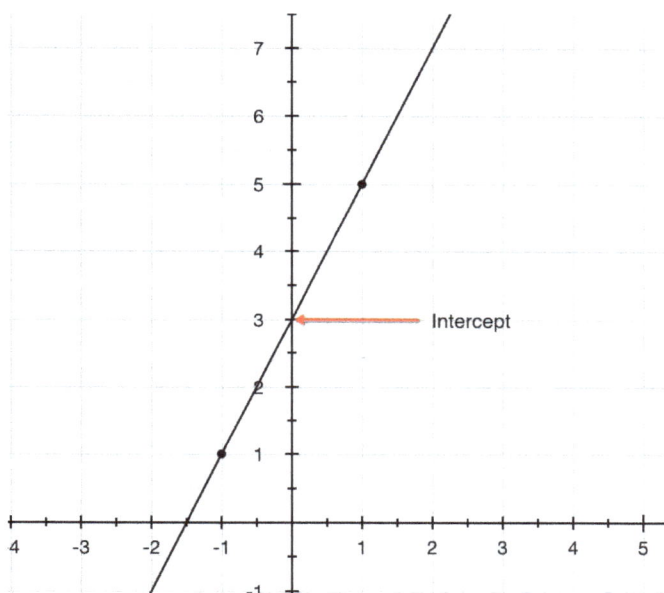

Gradient via two points

Let $A(x_1, y_1)$ and $B(x_2, y_2)$ be two points on a straight.

- The gradient of the line is given by $m = \dfrac{y_2 - y_1}{x_2 - x_1}$

- The length L of the line AB is given by $L = \sqrt{(x_2 - x_1)^2 + (y_2 - y_1)^2}$ using Pythagoras' theorem.

- The midpoint coordinates of the line AB are given by $\left(\dfrac{x_1 + x_2}{2}, \dfrac{y_1 + y_2}{2} \right)$

Example

(a) Determine the gradient of the straight line which passes through the points $A(1, 2)$ and $B(5, 4)$.

$$x_1 = 1, \quad y_1 = 2, \quad x_2 = 5, \quad y_2 = 4$$

$$m = \frac{y_2 - y_1}{x_2 - x_1} = \frac{4-2}{5-1} = \frac{2}{4} = \frac{1}{2}$$

(b) What is the equation of this line?

Equation of the straight line is $y = mx + c$, where we know that $m = \frac{1}{2}$ and that it passes through the point $A(1, 2)$, where $y = 2$ and $x = 1$. Thus

$2 = \frac{1}{2}1 + c$ so that $c = 2 - \frac{1}{2} = \frac{3}{2}$. Hence, the equation of the line is $y = \frac{1}{2}x + \frac{3}{2}$

(c) Plot this line.

> Notice the line passes through the points A and B and as expected.

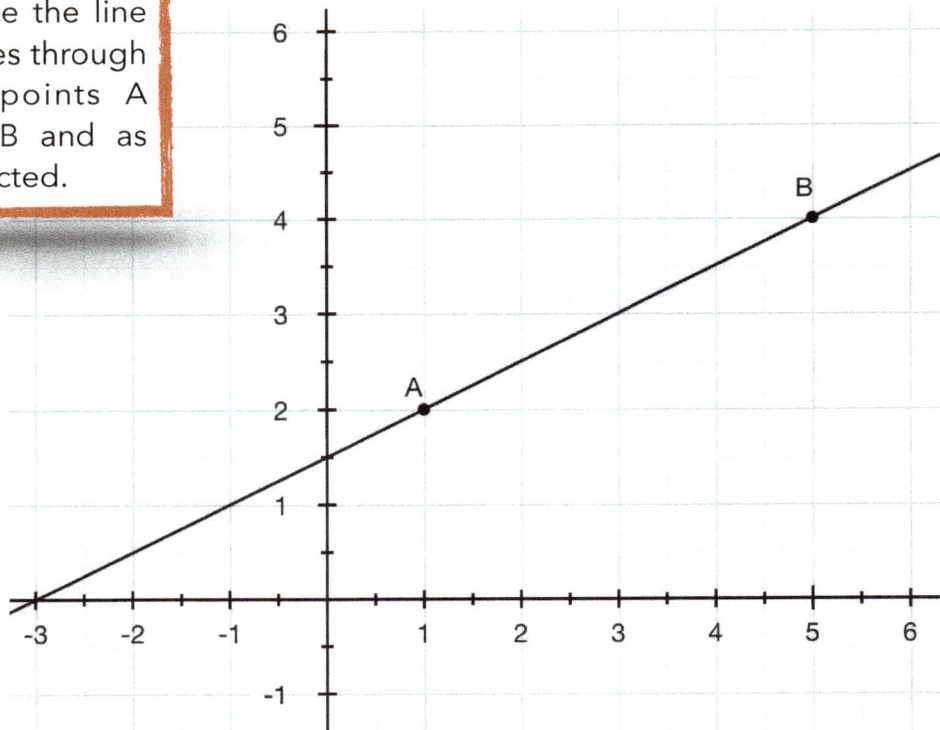

(d) What is the length of this line AB?

$$L = \sqrt{(x_2 - x_1)^2 + (y_2 - y_1)^2}$$
$$= \sqrt{(5-1)^2 + (4-2)^2}$$
$$= \sqrt{4^2 + 2^2} = \sqrt{20}$$

This is approximately 4.47 in length.

(e) What are the coordinates of the mid-point C between points A and B? Show this point on the line.

$$x_1 = 1, \; y_1 = 2, \; x_2 = 5, \; y_2 = 4$$

$$\left(\frac{x_1 + x_2}{2}, \frac{y_1 + y_2}{2} \right) = \left(\frac{6}{2}, \frac{6}{2} \right) = (3,3)$$

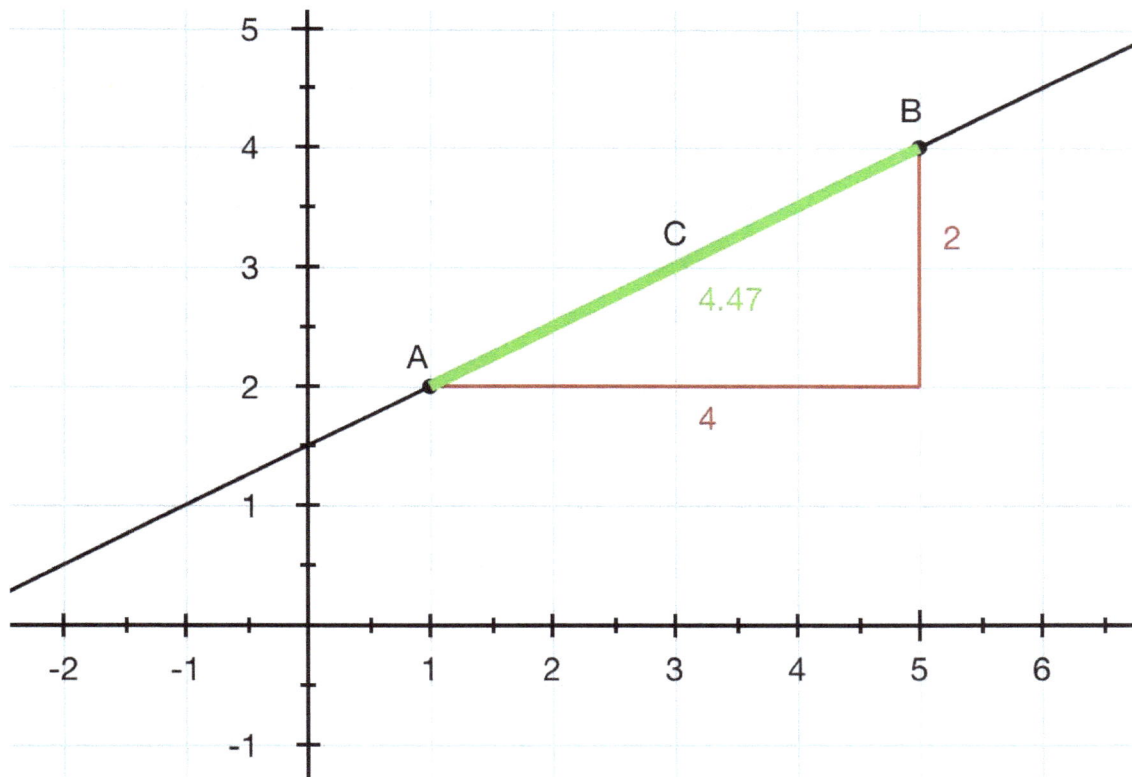

Two parallel straight lines have equal gradients.   However, if they are perpendicular, then the product of their gradients is equal to -1.

Examples

1.   State the equation of a line parallel to the line $y = 2x + 3$ with an intercept of 4 and show both lines on a graph.

For parallel lines, we require the same gradient. The line $y = 2x + 3$ has a gradient of 2 and intercept of 3. But we require an intercept of 4, thus we plot $y = 2x + 4$ as shown below.

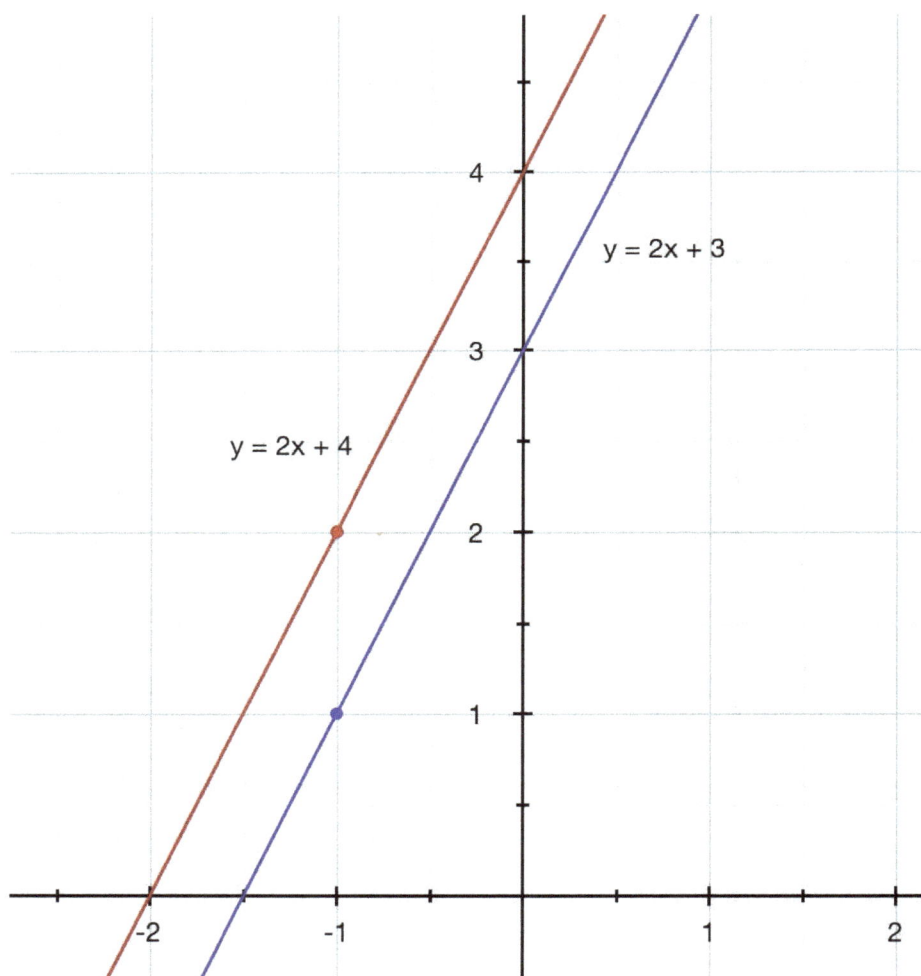

2.   State the equation of a line perpendicular to the $y = 2x + 3$ with the same intercept and show both lines on a graph.

For a perpendicular line, we require the product of their gradients be equal to -1. Let m be the gradient of the perpendicular line. Then

$$2m = -1$$

$$m = -\frac{1}{2}$$

For the same intercept of 3, the equation of the perpendicular line is $y = -\frac{1}{2}x + 3$ as shown below.

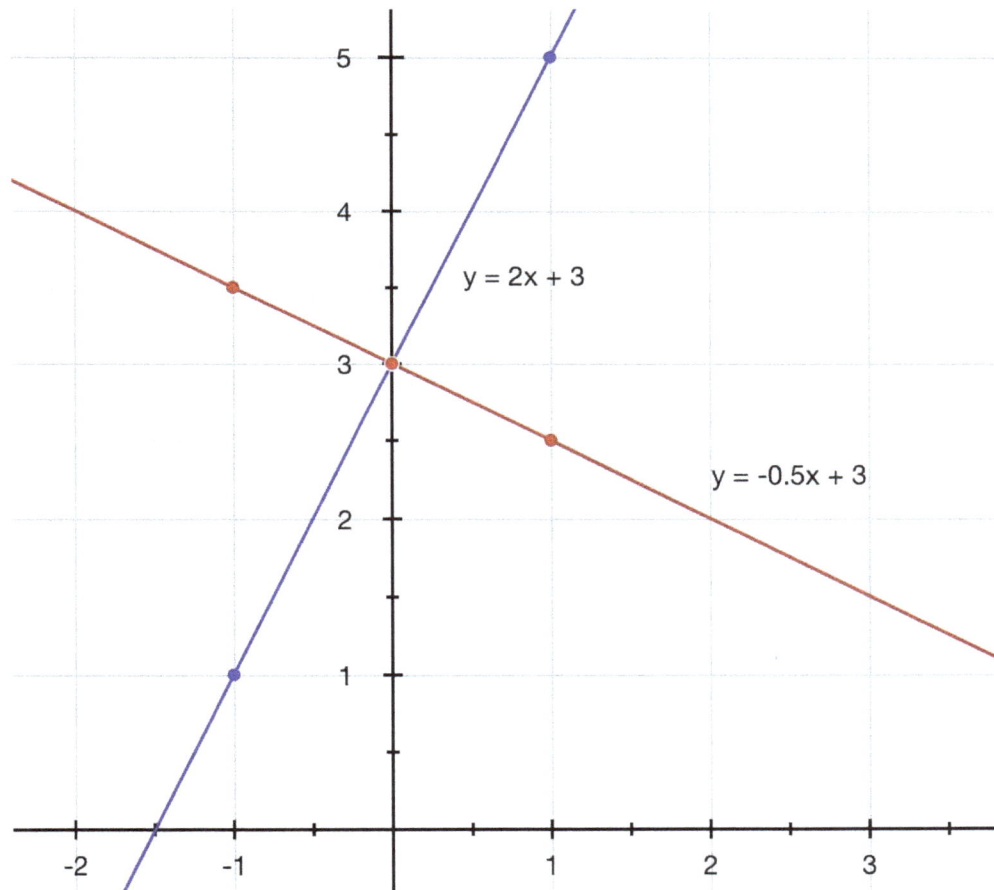

Substituting a numerical value into a function

- A function always come in the form $f(x)$, $g(x)$, $h(x)$, etc.
- A question may say find $f(2)$, which simply means to replace every x in the function with the number 2 and solve.

Example

1. Given $f(x) = x+3$, solve $f(4)$

$f(x) = x+3$

$f(4) = 4+3$

$f(4) = 7$

> The numerical value being substituted always replaces the term found in the function brackets and also the term found in the function equation

2. Given $h(x) = 2x-5$, solve $h(1)$

$h(x) = 2x-5$

$h(1) = 2(1)-5$

$h(1) = 2-5 = -3$

> The second function is **always** the function that is replacing the x-values and the first function is always the function being substituted into.

3. Given $g(x) = \dfrac{x+2}{x-1}$, solve $g(-3)$

$g(x) = \dfrac{x+2}{x-1}$

$g(-3) = \dfrac{-3+2}{-3-1}$

$g(-3) = \dfrac{-1}{-4} = \dfrac{1}{4}$

Combining two functions together to make one function

- A composite function usually looks like this $fg(x)$. This function is a combination of the $f(x)$ and $g(x)$ functions.

- In the $fg(x)$ function, the $f(x)$ function is first and the $g(x)$ function is second. You always substitute the entire **second** function for every x in the first function.

Example

1.   Given $g(x) = x + 3$ and $f(x) = x + 5$, determine $fg(x)$

Replace the x in $f(x)$ with the entire $g(x)$ function
$fg(x) = (x + 3) + 5$
$fg(x) = x + 8$

2.   Given $h(x) = 2x - 5$ and $g(x) = x + 2$, determine $hg(x)$ and hence solve $hg(2)$.

Replace the x in $h(x)$ with the entire $g(x)$ function
$hg(x) = 2(x + 2) - 5$
$hg(x) = 2x + 4 - 5$
$hg(x) = 2x - 1$

Hence
$hg(2) = 2(2) - 1 = 4 - 1 = 3$

3.   Given $f(x) = x + 4$ and $g(x) = \dfrac{x + 3}{x - 9}$, determine $gf(x)$

Replace the x in $g(x)$ with the entire $f(x)$ function

$gf(x) = \dfrac{(x + 4) + 3}{(x + 4) - 9}$

$gf(x) = \dfrac{x + 4 + 3}{x + 4 - 9}$

$gf(x) = \dfrac{x + 7}{x - 5}$

Derive the inverse of a function

The inverse function is basically a function that undoes what the original function has done. To determine the inverse function of $f(x) = px + c$

Step 1   let $f(x) = y$

$y = px + c$

Step 2   Interchange x and y (switch every y with x and every x with a y)

$x = py + c$

Step 3   Make y the subject of the equation in the previous step

$x = py + c$

$x - c = py$

$\dfrac{x - c}{p} = y$

$y = \dfrac{x - c}{p}$

> Always follow these four steps in order to achieve the inverse.

Step 4   Replace the y with the inverse of $f(x)$ symbol $f^{-1}(x)$

$f^{-1}(x) = \dfrac{x - c}{p}$

Example

1.   Given $f(x) = x - 7$, find $f^{-1}(x)$ and solve $f^{-1}(3)$

$y = x - 7$
$x = y - 7$
$x + 7 = y$
$y = x + 7$
$f^{-1}(x) = x + 7$
Hence
$f^{-1}(3) = 3 + 7 = 10$

2.   Given $g(x) = 3x - 2$, find $g^{-1}(x)$

$y = 3x - 2$
$x = 3y - 2$
$x + 2 = 3y$
$\dfrac{x + 2}{3} = y$ or $y = \dfrac{x + 2}{3}$

$g^{-1}(x) = \dfrac{x + 2}{3}$

3.   If $h(x) = \dfrac{2x+4}{x-5}$ then determine $h^{-1}(x)$ and hence $h^{-1}(3)$

$y = \dfrac{2x+4}{x-5}$

$x = \dfrac{2y+4}{y-5}$

Cross multiply the entire denominator
$x(y-5) = 2y+4$

Multiply out the bracket
$xy - 5x = 2y + 4$

Put all the y terms together and everything else on the other side
$xy - 2y = 4 + 5x$

Factorize the left hand side for y
$y(x-2) = 4 + 5x$

Make y the subject
$y = \dfrac{4+5x}{x-2}$

$h^{-1}(x) = \dfrac{4+5x}{x-2}$

Hence
$h^{-1}(3) = \dfrac{4+5(3)}{3-2} = \dfrac{4+15}{3-2} = \dfrac{19}{1} = 19$

Use the relationship (fg)^-1(x) =g^-1f^-1(x)

The equation $\left(fg\right)^{-1}(x) = g^{-1}f^{-1}(x)$ can be read in both directions.

Examples

1.   Show that $\left(fg\right)^{-1}(x) = g^{-1}f^{-1}(x)$ is true if $f(x) = x+4$ and $g(x) = x+6$

Step 1   Determine $\left(fg\right)^{-1}(x)$

$$fg(x) = (x+6)+4$$
$$fg(x) = x+6+4$$
$$fg(x) = x+10$$

Let $y = fg(x)$
$$y = x+10$$
$$x = y+10$$
$$x-10 = y \quad or \quad y = x-10$$
$$(fg)^{-1}(x) = x-10$$

Step 2   Determine $g^{-1}f^{-1}(x)$
$$g(x) = x+6$$
$$y = x+6$$
$$x = y+6$$
$$x-6 = y \quad or \quad y = x-6$$
$$g^{-1}(x) = x-6$$

$$f(x) = x+4$$
$$y = x+4$$
$$x = y+4$$
$$x-4 = y \quad or \quad y = x-4$$
$$f^{-1}(x) = x-4$$

Now replace the x in $g^{-1}(x)$ with entire $f^{-1}(x)$

Therefore
$$g^{-1}f^{-1}(x) = (x-4)-6$$
$$g^{-1}f^{-1}(x) = x-4-6 = x-10$$

Step 3   Therefore $(fg)^{-1}(x) = g^{-1}f^{-1}(x)$ since $x-10 = x-10$

2.   If $f(x) = x-2$ and $h(x) = x-1$,   determine $(fh)^{-1}(x)$, hence calculate $h^{-1}f^{-1}(2)$

$fh(x) = (x-1) - 2$

$fh(x) = x - 1 - 2$

$fh(x) = x - 3$

$y = x - 3$

$x = y - 3$

$x + 3 = y \ \ or \ \ y = x + 3$

$(fh)^{-1}(x) = x + 3$

Hence calculate $h^{-1}f^{-1}(2)$

Since we know that $(fh)^{-1}(x) = h^{-1}f^{-1}(x)$ then

$h^{-1}f^{-1}(x) = x + 3$

$h^{-1}f^{-1}(2) = 2 + 3 = 5$

Axis of symmetry, maximum or minimum value of a quadratic function

- The minimum/maximum point on the graph is the point where the graph turns. It is usually denoted as $ymin$ or $ymax$ respectively.

- The axis of symmetry, which is usually written as $x = d$, where d is an integer, is a line that cuts the curve into two equal halves.

- a is the coefficient in front of the x^2 term.

- $h = \dfrac{b}{2a}$

- $k = \dfrac{4ac - b^2}{4a}$

- axis of symmetry is x=-h
- minimum value and the maximum value = k
- If the coefficient of x² is positive, then the function has a minimum value.
- If the coefficient of x² is negative, then the function has a maximum value.

Examples

1. Write the following quadratic equation $x^2 + 5x + 4$ in the form $a(x+h)^2 + k$

$x^2 + 5x + 4$ can also be represented as $ax^2 + bx + c$ where the integers $a = 1$, $b = 5$ and $c = 4$.

$$h = \frac{b}{2a} = \frac{5}{2 \times 1} = \frac{5}{2}$$

$$k = \frac{4ac - b^2}{4a} = \frac{4 \times 1 \times 4 - 5^2}{4 \times 1}$$

$$\frac{16 - 25}{4} = \frac{-9}{4}$$

Therefore in the form $a(x+h)^2 + k$

$$1\left(x + \frac{5}{2}\right)^2 - \frac{9}{4}$$

2.   For the quadratic function $2(x+5)^2 - 9$, find the equation of the axis of symmetry and the minimum value of the function.

Based on the above quadratic function we can see that $h = 5$ and $k = -9$. We know that the equation of the axis of symmetry is $x = -h$ and the minimum value, $y_{min} = k$.

Therefore the equation of the axis of symmetry is
$x = -h$ and we know that $h = 5$
$x = -(5)$
$x = -5$

The minimum value y_{min} is
$y_{min} = k$ and since $k = -9$
$y_{min} = -9$

3.   Determine by method of completing the squares, the equation of the axis of symmetry and the minimum value of the quadratic equation

$$f(x) = -2x^2 + 4x + 1$$
$$a = -2, b = 4, c = 1$$

Using the completing the squares method we know that $h = \dfrac{b}{2a}$ and $k = \dfrac{4ac - b^2}{4a}$

$$h = \frac{b}{2a} = \frac{4}{2 \times -2} = \frac{4}{-4} = -1$$

$$k = \frac{4ac - b^2}{4a} = \frac{4 \times -2 \times 1 - (4)^2}{4 \times -2}$$

$$= \frac{-8 - 16}{-8} = \frac{-24}{-8} = 3$$

Therefore the equation of the axis of symmetry is
$x = -h$ and since $h = -1$

$x = -(-1)$

$x = 1$

The maximum value y_{max} is
$y_{max} = k$ and since $k = 3$

$y_{max} = 3$

Sketch a graph of quadratic function

To sketch a graph of a quadratic function expressed as a $a(x + h)^2 + k$, you need to know that,

- the axis of symmetry is $x = -h$
- the minimum or maximum valve $y_{min} = k$ or $y_{max} = k$
- the roots or x intercepts are found via $y = 0$
- similarly the y intercept is found via $x = 0$
- the turning point of the graph is $(-h, k)$

Examples

1.   Given the function $y = x^2 + 2x - 3$, sketch the graph using the method of completing the squares or $a(x + h)^2 + k$

Since we have $y = x^2 + 2x - 3$ then we know $a = 1, b = 2, c = -3$

i) Determine the axis of symmetry, which is $x = -h$

$$h = \frac{b}{2a} = \frac{2}{2 \times 1} = \frac{2}{2} = 1$$
$$x = -h = -(1) = -1$$

ii) The value of the coefficient a is positive so therefore we will have a minimum value y_{min}

$$k = \frac{4ac - b^2}{4a} = \frac{4 \times 1 \times -3 - (2)^2}{4 \times 1} = \frac{-12 - 4}{4} = \frac{-16}{4} = -4$$
$$y_{min} = k$$
$$y_{min} = -4$$

Therefore the quadratics equation in the form or $a(x+h)^2 + k$

is $y = 1(x+1)^2 - 4$

iii) To find the roots or x-intercepts we make y or $f(x) = 0$

$$1(x+1)^2 - 4 = y$$
$$1(x+1)^2 - 4 = 0$$
$$1(x+1)^2 = 0 + 4$$
$$1(x+1)^2 = 4$$
$$(x+1)^2 = \frac{4}{1}$$
$$(x+1)^2 = 4$$
$$(x+1) = \pm\sqrt{4}$$
$$x + 1 = \pm 2$$
$$x = \pm 2 - 1$$
$$x = +2 - 1 \qquad x = -2 - 1$$
$$x = 1 \qquad\qquad x = -3$$

Or as in this case, if you are given the complete quadratic equation, you can simply factorize and solve to get the roots.

$y = x^2 + 2x - 3$

$x^2 + 2x - 3 = 0$

$(x+3)(x-1) = 0$

$(x+3) = 0$ and $(x-1) = 0$

$x = 0 - 3$ $x = 0 + 1$

$x = -3$ $x = 1$

> This method can **only** be used if you are given the full quadratic equation.

iv) To determine the y-intercept, set $x = 0$

$y = 1(x+1)^2 - 4$

$y = 1(0+1)^2 - 4$

$y = 1(1)^2 - 4$

$y = 1 - 4$

$y = -3$

so the y intercept is -3 and the y intercept coordinate is (0,-3)

Or equivalently, if the full quadratic equation is given then the y intercept is always c, so in this case using the equation $y = x^2 + 2x - 3$, the y intercept is -3 and the y intercept coordinate = (0,c) = (0,-3).

v) The turning point of the graph is (-h,k), hence the turning point of this graph is (-1,-4).

vi) Sketch the graph of $y = 1(x+1)^2 - 4$ or $y = x^2 + 2x - 3$

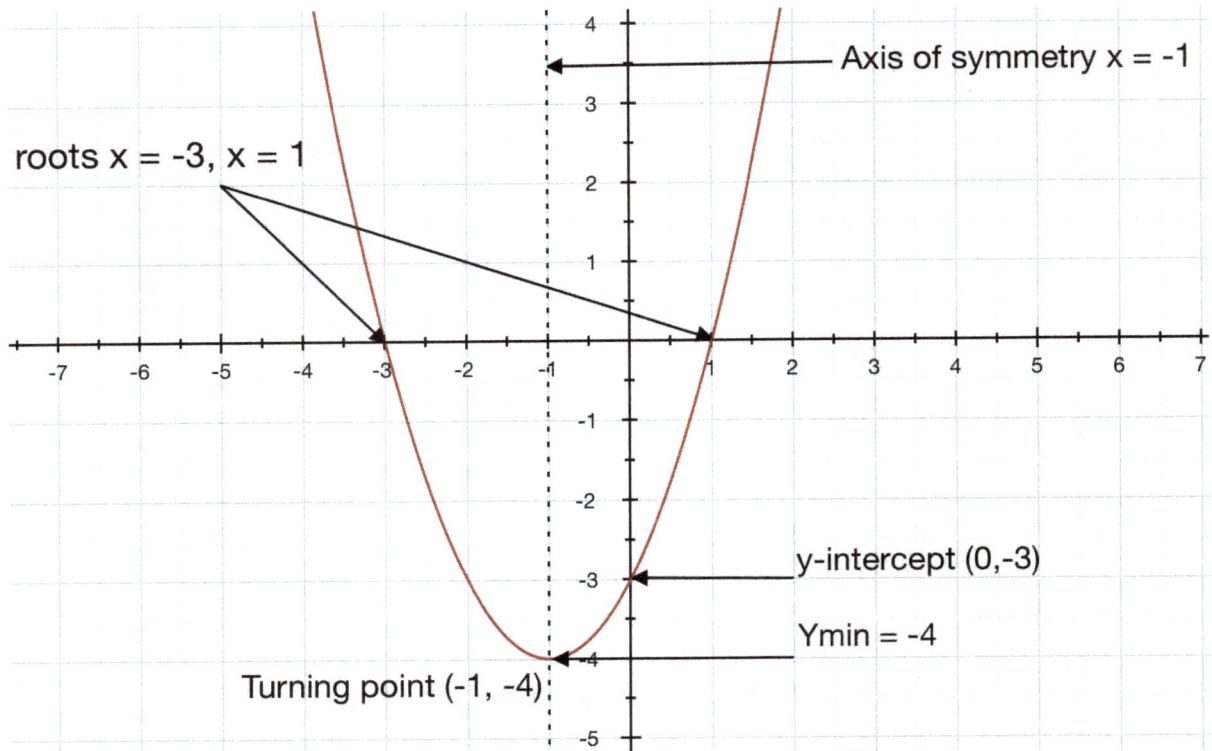

Axis of symmetry x = -1

roots x = -3, x = 1

y-intercept (0,-3)

Ymin = -4

Turning point (-1, -4)

2.   Sketch the graph of $f(x) = 1(x+2)^2 - 1$ and find its roots.

We know that h = 2 and k = -1 based on the equation being used.

i) The axis of symmetry $x = -h$
$h = 2$
$x = -h = -(2) = -2$

ii) The value of the coefficient a is positive so therefore we will have a minimum value y_{min}
$k = -1$
$y_{min} = k$
$y_{min} = -1$

iii) To find the roots or x-intercepts, we set $y = 0$.

$$1(x+2)^2 - 1 = y$$
$$1(x+2)^2 - 1 = 0$$
$$1(x+2)^2 = 0 + 1$$
$$1(x+2)^2 = 1$$
$$(x+2)^2 = \frac{1}{1}$$
$$(x+2)^2 = 1$$
$$(x+2) = \pm\sqrt{1}$$
$$x + 2 = \pm 1$$
$$x = \pm 1 - 2$$
$$x = +1 - 2$$

$$x = +1 - 2 \qquad and \qquad x = -1 - 2$$
$$x = -1 \qquad\qquad\qquad\qquad x = -3$$

iv) To determine the y-intercept set $x = 0$.

$$y = 1(x+2)^2 - 1$$
$$y = 1(0+2)^2 - 1$$
$$y = 1(2)^2 - 1$$
$$y = 4 - 1$$
$$y = 3$$

so the y intercept is 3 and the y-intercept coordinate is (0,3)

vi) The turning point of the graph is (-h,k), so the turning point of this graph is (-2,-1).

vii) Sketch the graph of $f(x) = 1(x+2)^2 - 1$

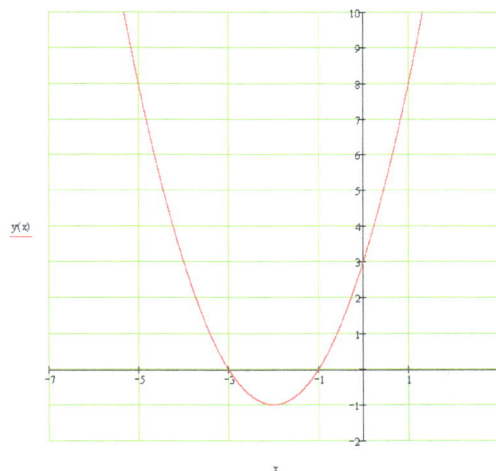

Drawing graphs of the type y = ax^n where n is a constant

Illustrated by the following example.

Example

1.   Draw the graph of the type $y = x^3$ for the domain $-2 \leq x \leq 2$

x	-2	-1	0	1	2
$y=x^3$	-8	-1	0	1	8

For the domain $-2 \leq x \leq 2$, the graph of $y = x^3$ is shown below.

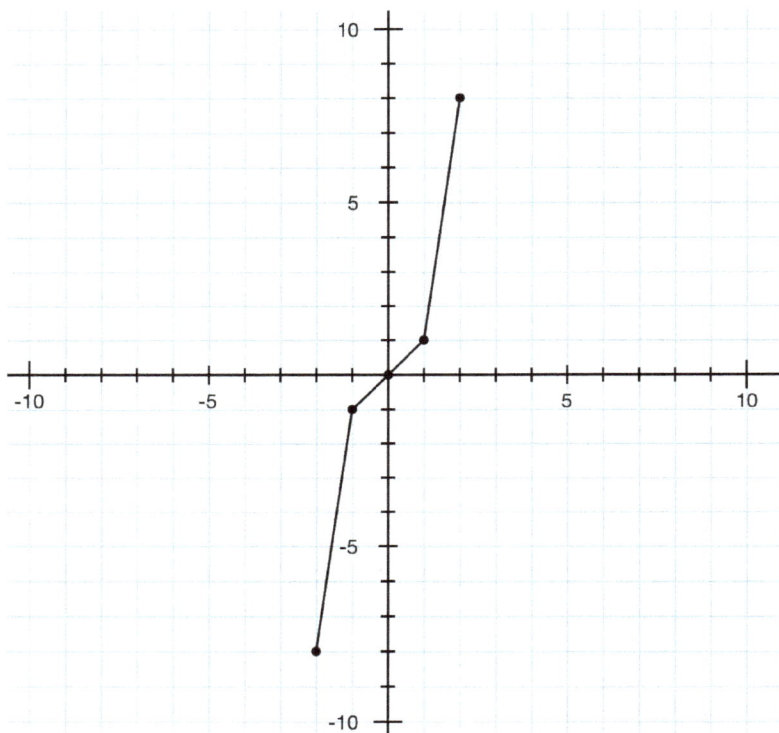

2)   Draw the graph of the type $y = 2x^{-1}$ for the domain $-3 \leq x \leq 3$

x	-3	-2	-1	0	1	2	3
x^{-1}	-0.3	-0.5	-1	undefined	1	0.5	0.3
$y=2x^{-1}$	0.6	-1	-2	undefined	2	1	0.6

For the domain $-3 \leq x \leq 3$, the graph of $y = 2x^{-1}$ is shown below.

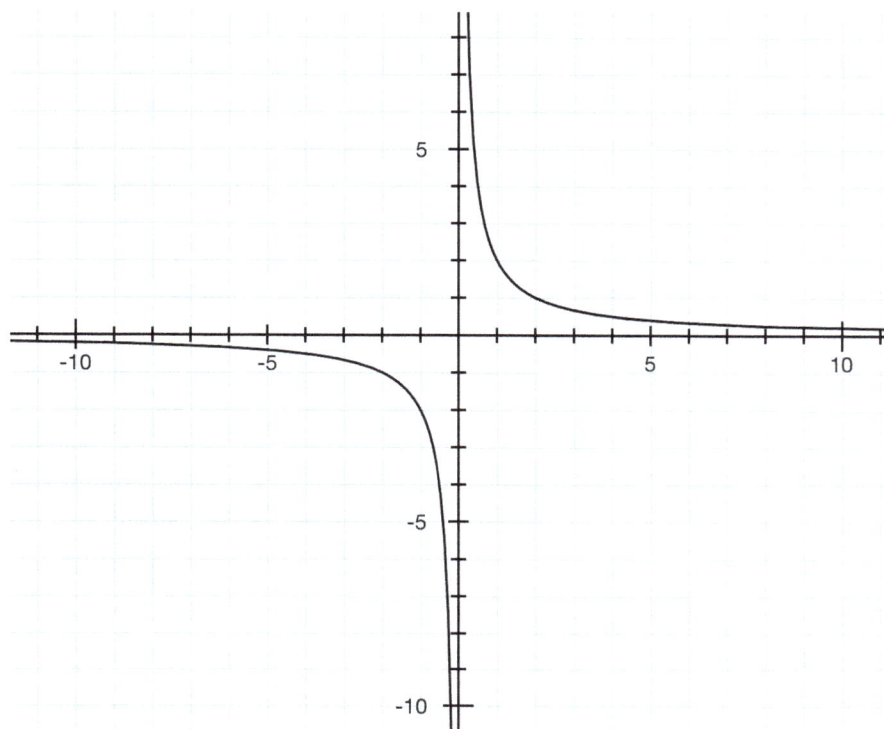

3)   Draw the graph of the type $f(x) = x^{-2}$ for the domain $-3 \leq x \leq 3$

x	-3	-2	-1	0	1	2	3
y=x⁻¹	0.1	0.3	1	undefined	1	0.3	0.1

For the domain $-3 \leq x \leq 3$, the graph of $f(x) = x^{-2}$ is shown below.

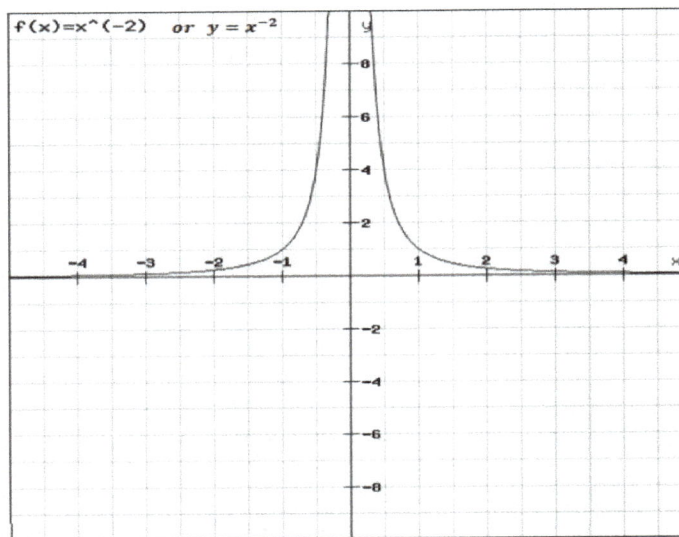

f(x)=x^(-2)   or $y = x^{-2}$

Draw and interpret distance time graphs and speed time graphs

- **Distance** is the amount of space found between two objects or points. Its given by the speed at which an object is travelling multiplied by the take it takes to reach its destination.

- **Time** is how long it takes an object to travel the distance between two points.

- **Speed** is the total distance travelled divided by the total time taken.

- **Acceleration magnitude** is the change in velocity (displacement over time) divided by change in time.

> speed=distance/time.
> distance=speed x time
> time=distance/speed

Examples

1.   Calculate the speed if an object travels 20 km in 2 hours.

$$Speed = \frac{distance}{time} = \frac{20\ km}{2\ hours} = 10\ km\ per\ hour \quad \text{or } 10\ kmh^{-1}$$

2.   Calculate the distance if a boat is travelling at a speed of 15 ms^{-1} for 2 hours.

The speed is in ms^{-1} therefore we must change our time which is in hours to seconds

2 hours = 2×60 seconds=120 seconds

$Distance = speed \times time = 15 \ ms^{-1} \times 120 \ seconds = \ 1800 \ m$

3.   A man journeys 2000 km to his hometown at a speed of 5 kmh^{-1}. Calculate the time it takes him to reach home

$$Time = \frac{distance}{speed} = \frac{2000 \ km}{5 \ kmh^{-1}} = 400 \ hours$$

4.   A cyclist rides for 200 km at a speed of $100 kmh^{-1}$, 100 km at a speed of $25 kmh^{-1}$ and a further 200km at a speed of $40 kmh^{-1}$.

(a) Draw a distance time graph to represent the journey and use the graph to determine, (b) the total distance travelled, (c) the total time taken and (d) the average speed for the journey of the cyclist.

In order to draw a distance time graph you need both distance and time. However the information given to us only gives us distance and speed. So we must determine the time taken for each segment of the journey in order to plot a graph of distance vs time.

First segment   distance travelled = 200 km, speed = $100 kmh^{-1}$
Therefore the time taken for first segment

$$Time = \frac{distance}{speed} = \frac{200 \ km}{100 \ kmh^{-1}} = 2 \ hours$$

Second segment   distance travelled = 100 km, speed = $25 \ kmh^{-1}$
Therefore the time taken for second segment

$$Time = \frac{distance}{speed} = \frac{100 \ km}{25 \ kmh^{-1}} = 4 \ hours$$

Third segment   distance travelled = 200 km, speed = $40 \ kmh^{-1}$
Therefore the time taken for second segment,

$$Time = \frac{distance}{speed} = \frac{200 \ km}{40 \ kmh^{-1}} = 5 \ hours$$

The Distance vs Time table of points

Distance (Km)	200	100	200
Time (hours)	2	4	5

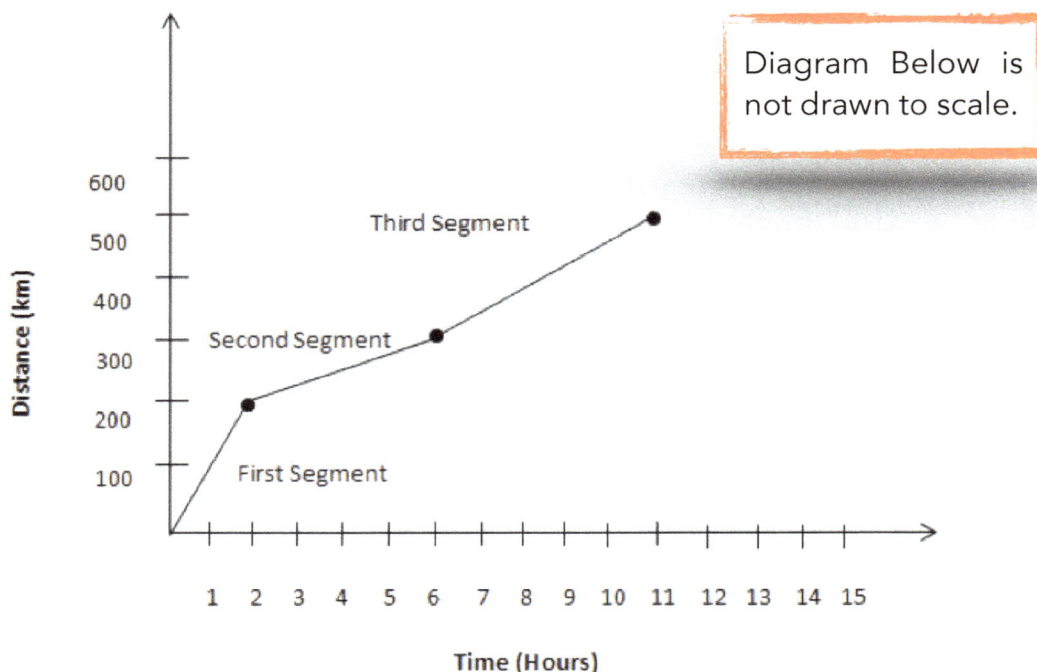

(b) The total distance travelled = 200 km + 100 km + 200 km = 500 km

Using the graph we see the journey ends at the 500 km mark, which also indicates the total distance travelled.

(c) The time taken = $2\ hours + 4\ hours + 5\ hours = 11\ hours$

Using the graph we see the journey ends at the 11 hour mark, which indicates that the total time taken is 11 hours.

(d) $Average\ speed\ for\ entire\ journey = \dfrac{total\ distance\ travelled}{total\ time\ taken}$

$$= \frac{500\ km}{11\ hours} = 45.45\ kmh^{-1}\ (2\text{dp})$$

Speed-time graphs and velocity-time graphs

- **Displacement** is defined as the change in position of an object in a particular direction

- **Velocity** of an object is defines as the rate in which an object changes its position divided by the time taken to change.

- **Time** is defined as how long it takes an object to travel the distance between two points

- **Acceleration** is defined as the rate at which velocity changes divided by the time taken to make that change.

- **Retardation /deceleration** means to slow down and   is defined as negative acceleration

- $displacement = velocity \times time$

- $velocity = \dfrac{displacement}{time}$ or $velocity = acceleration \times time$

- $time = \dfrac{displacement}{velocity}$

- $acceleration = \dfrac{velocity\ change}{time\ taken} = \dfrac{v_2 - v_1}{t_2 - t_1}$

Examples

1.   A bus starting from rest (no velocity or time) accelerates to a velocity 10 10 ms^{-1} or 10 m/s in 2 seconds. Calculate the average acceleration of the car.

$v_1 = 0\ ms^{-1}$ $\qquad\qquad$ $t_1 = 0\ seconds$

$v_2 = 10\ ms^{-1}$ $\qquad\qquad$ $t_2 = 2\ seconds$

$$acceleration = \frac{velocity\ change}{time\ taken} = \frac{v_2 - v_1}{t_2 - t_1}$$

$$= \frac{(10-0)ms^{-1}}{(2-0)secs} = \frac{10ms^{-1}}{2\ secs} = 5ms^{-2} \text{ or } 5m/s^2$$

2.   A cyclists starts at rest and moves with a uniform acceleration of $25\ m/s^2$ for 4 seconds. Calculate the velocity of the cyclist.

$velocity = acceleration \times time$

$= 25\ m/s^2 \times 4\ seconds$

$= 100 m/s$

3.   A motorists accelerates from rest uniformly to a speed of $40\ m/s$ in 5 seconds. This speed is maintained for another 15 seconds and then applies brakes and decelerates uniformly to rest in 5 seconds.

(a) Draw a velocity- time graph to show the entire journey

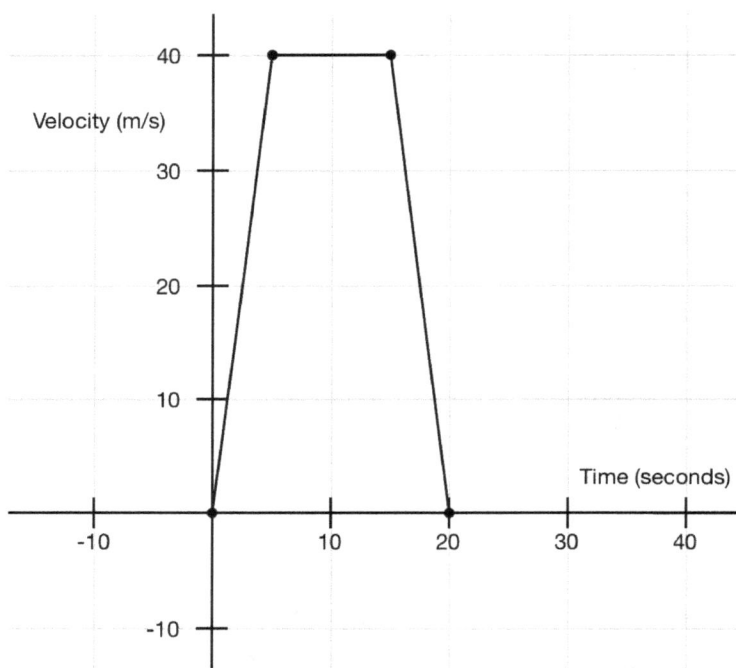

(b) Calculate the acceleration in the first 5 seconds

$$acceleration = \frac{velocity\ change}{time\ taken} = \frac{v_2 - v_1}{t_2 - t_1}$$

$$= \frac{(40-0)ms^{-1}}{(5-0)secs} = \frac{40\ ms^{-1}}{5\ secs} = 8ms^{-2}$$

(c) Calculate the retardation in the last 5 seconds

$$retardation = \frac{v_2 - v_1}{t_2 - t_1}$$

$$= \frac{(0-40)ms^{-1}}{(20-15)secs} = \frac{-40\ ms^{-1}}{5\ secs} = -8ms^{-2}$$

(d) Calculate the total length of the journey

The total length of the journey = Area of trapezium

$$= \frac{1}{2}(a+b)h$$

$$= \frac{1}{2}(10+20)s \times 40\ m/s$$

$$= 600\ metres$$

(e) Calculate the average speed of the journey

$$Average\ speed\ for\ entire\ journey = \frac{total\ distance\ travelled}{total\ time\ taken}$$

$$= \frac{600\ m}{20\ seconds} = 30\ ms^{-1}$$

4.   A motorists accelerates from rest uniformly to a speed of 30 m/s in 5 seconds. This speed is maintained for another 20 seconds and then applies brakes and decelerates uniformly to rest in 10 seconds.

(a) Draw a velocity- time graph to show the entire journey.

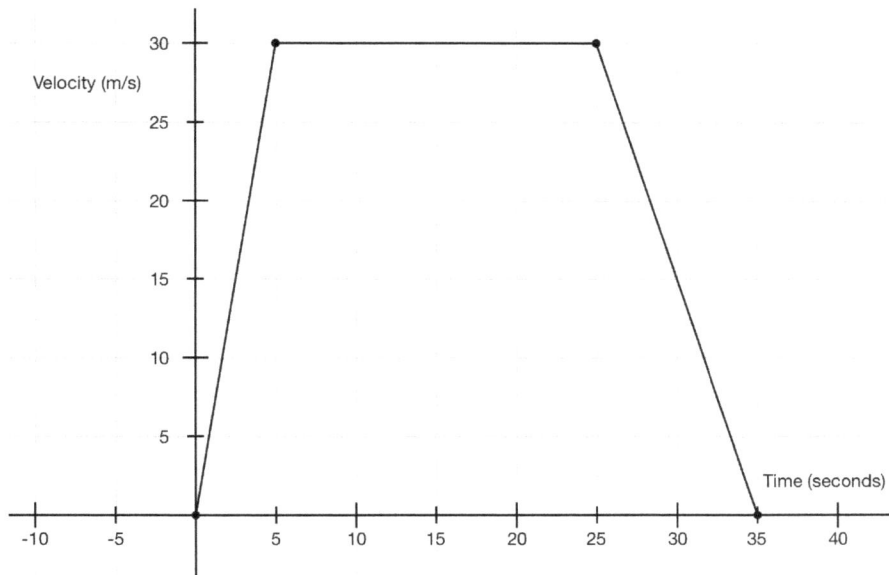

(b) Calculate the distance travelled in the first 5 seconds.

Distance in first 5 seconds = Area of triangle $=\dfrac{1}{2}bh$

$=\dfrac{1}{2}(5\ s\times 30m/s)=75\ m$

(c) Average speed of the journey.

Total distance travelled = Area of trapezium$=\dfrac{1}{2}(a+b)h$

$=\dfrac{1}{2}(20+35)s\times 30\ m/s$

$=825\ metres$

$Average\ speed\ for\ entire\ journey =\dfrac{total\ distance\ travelled}{total\ time\ taken}$

$=\dfrac{825\ m}{35\ seconds}=23.6\ ms^{-1}$ (1 d.p)

Questions

Relations, Functions and Graphs

Kemar Trotman

The Lodge School
Barbados

Phillip Clarke

Parkinson Memorial
School, Barbados

QUESTIONS

[1] Given the function $f(x) = x + 2$, calculate the value of $f(2)$.

[2] Given the function $g(x) = ((x + 7) \div (x + 2))$, calculate $g(3)$.

[3] Given the function $h(x) = 5x + 2$, determine $h^{-1}(x)$.

[4] Given the function $f(x) = 2x + 3$ and $g(x) = 4x + 7$, calculate the value of the function $fg(x)$.

> ‣ Video solutions to the questions can be viewed via the App "**CTS Maths**" in the Apple and Google Play stores.

[5] Draw the graph of the quadratic equation $f(x) = x^2 + 2x - 3$ for the domain $-5 \leq x \leq 3$.

[6] Given the quadratic equation $f(x) = x^2 + 5x + 6$, determine the roots of this quadratic function $f(x)$.

[7] Given the function $f(x) = x^2 + 2x - 8$ with domain $-5 \leq x \leq 3$, determine the maximum or minimum value of $f(x)$.

[8] Given the quadratic equation $f(x) = x^2 - 2 \times + 8$ with domain $-5 \leq x \leq 3$, determine the maximum or minimum value of the quadratic equation $f(x)$.

[9] (a) Draw the axis of symmetry on the graph shown below.
     (b) State the equation of the axis of symmetry.

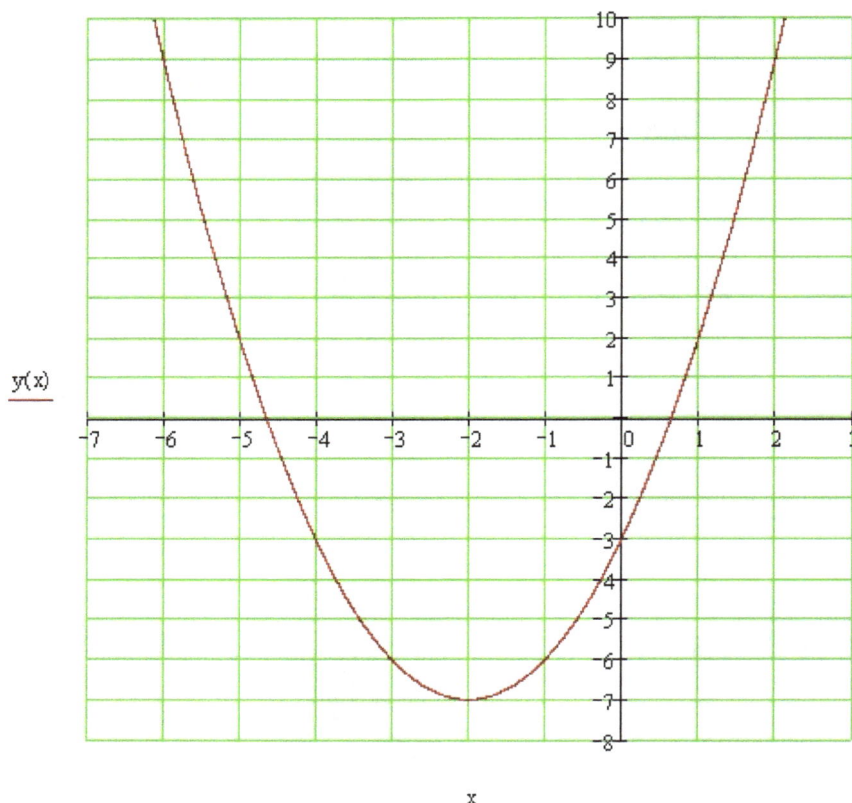

[10] Write the quadratic equation $2x^2 - 5 \times -3$ in the form $a(x + h)^2 + k$.

[11] Find the inverse of the function $f(x) = \dfrac{3x-2}{4x+7}$

[12] The equation of the line AB is $3y = 4x+1$. Find the equation of the line XY that is parallel to AB and passes through the origin.

[13] The equation of the line PQ is $y = 5x-3$. Find the equation of the line perpendicular to PQ, which is called LM, if it passes through the point (2,7).

[14] Find the length of the line segment which joins the two points A(-1,3) and B(4,5).

[15] (a) Find the coordinates of the midpoint X which lies between the points A(-6,7) and B(4,0).
(b) Find the equation of the line which passes through the points A and B.

[16] Solve the following pair of simultaneous equations graphically where   .
$0 \le x \le 5$ $2y = 4x-1$ $y = 2-3x$

[17] Draw the graph of $y = 2x^{-3}$ for $-2 \le x \le 2$

[18] Draw the graph of $y = \dfrac{1}{2}x^{-1}$ for $-2 \le x \le 2$

[19] A car travels for 300 km at 100km/h, 150 km at a speed of 75km/h and 50 km at a speed of 50km/h.

(a)   Draw a distance time graph to show the above information.
(b)   Find the average speed for the entire journey.

[20] A car is at a junction that leads onto a highway. When the road is clear it accelerates from rest uniformly to a speed of 30 m/s in 4 seconds. After 11 seconds a roundabout approaches, so the driver brakes and decelerates uniformly to a stop in 5 seconds.
i.   Draw a Velocity Time graph.
ii.   Calculate the acceleration in the first 4s.
iii.   Calculate the deceleration.
iv.   Calculate the distance travelled.

Chapter 9
Statistics
Stuart Mayers
Combermere School
Barbados

Statistics can help us analyse and understand various aspects of what is happening in the world around us. It is one of the most applied branches of Mathematics, with several uses and implications for our everyday lives. With statistics we can study what has happened in the past and use this information to predict what may happen in the future. Weather forecasts, medical studies and political polls are some examples of common uses of Statistics.

Types of data

Data can be classified in different ways.

- **Qualitative** (categorical) data is information about different groups or categories, such as colors or names. The differences between the groups shows no rank or order (red is not better than blue or vice versa).

- **Quantitative** (numerical) data is any form of data that can be classified based on a number value. There are two types of quantitative data; discrete and continuous data.

- **Discrete** data is any type of information that can be counted exactly. For example, the number of cars passing a particular point in half an hour, or the number of children in your class.

- **Continuous** data is numerical data that has interval values or lie within a range. This information usually comes about through some form of measurement. Weights, heights, lengths and time are all examples of continuous data.

Measures of central tendency

The arithmetic mean of a set of data is calculated using

$$arithmetic\ mean = \frac{sum\ of\ the\ values}{number\ of\ values}$$

- The arithmetic mean is very often referred to as "the average"
- The mode is the value that occurs the most often.
- The median of a set of data is the value in the middle when the values are arranged in ascending or descending order.
- When the number of items is large $(n \geq 30)$, simply divide the number of items in two. The value in that position is the median.
- When the number of items is small $(n < 30)$, we can use the following equation to assist in finding the median.

$$median = \left(\frac{n+1}{2}\right)^{th} value$$

where n, represents the total number of items.

Example

1.   Find the mean, mode and median for the data set $(1, 1, 7, 8, 8, 8, 9)$

$$arithmetic\ mean = \frac{sum\ of\ the\ values}{number\ of\ values}$$

$$= \frac{1+1+7+8+8+8+9}{7} = \frac{42}{7} = 6$$

$$mode = 8\ (there\ are\ more\ 8s\ than\ any\ other\ number)$$

$$median = \left(\frac{n+1}{2}\right)^{th} value$$

$$= \left(\frac{7+1}{2}\right)^{th} value$$

$$= \left(\frac{8}{2}\right)^{th} = 4^{th} value$$

The fourth value is 8. Therefore the median is 8.

2.   Find the mean, mode and median of the data set (7, 3, 5, 2, 7, 7, 9, 7, 2, 4)
First arrange the data in order.   2, 2, 3, 4, 5, 7, 7, 7, 7, 9

$$mean = \frac{sum\ of\ the\ values}{number\ of\ values}$$

$$= \frac{2+2+3+4+5+7+7+7+7+9}{10} = \frac{53}{10} = 5.3$$

$$mode = 7$$

$$median = \frac{n+1}{2}^{th} value$$

$$= \frac{10+1}{2}^{th} value$$

$$= 5.5^{th} value$$

> When the number of items is even, the median will lie between the two middle numbers in the data set. This is why it is necessary to find the mean of these two numbers.

The 5.5*th* value is between the fifth and sixth value. Here the fifth and sixth values are 5 and 7 respectively. Therefore the median is the mean of these two numbers.

$$median = \frac{5+7}{2} = 6$$

The median of this set of data is 6.

Frequency tables

• It often makes sense to place raw data into a frequency table. This can make the information easier to compare and analyze. Frequency tables can be used for both categorical and numerical data.

• In a frequency table the different categories are listed against the respective frequencies (the number of times each occurs).

Examples

1.   A set of fifty students were asked what was their favourite colour. The following table shows that data. How many students were interviewed?

COLOR	RED	BLUE	YELLOW	GREEN
FREQUENCY	14	21	10	5

Total number of students = sum of the frequencies = 14 + 21 + 10 + 5 = 50 students.

The sum of the frequencies is equal to the total number of observations (fifty in this case).

2.   The following is a sample of the marks from a group of students on an English test. Form a frequency table from the set of raw data:

42, 41, 42, 43, 45, 41, 42, 42, 40, 44, 44, 42, 41, 43, 44

MARK	TALLY	FREQUENCY
40	I	1
41	III	3
42	IIII	5
43	II	2
44	III	3
45	I	1

$$\sum f = 15$$

$\sum f$ means the sum of frequencies

Bar charts

A bar chart is a graphical representation of a frequency table for ungrouped discrete data. Each bar represents a category or value, and the height of each bar represents its frequency. Each of the bars should be the same distance apart and should be equal in width.

Example

Using the set of data from the previous example, form a bar chart to show the given information.

MARK	40	41	42	43	44	45
FREQUENCY	1	3	5	2	3	1

Bar chart showing the frequency of the marks on an English test

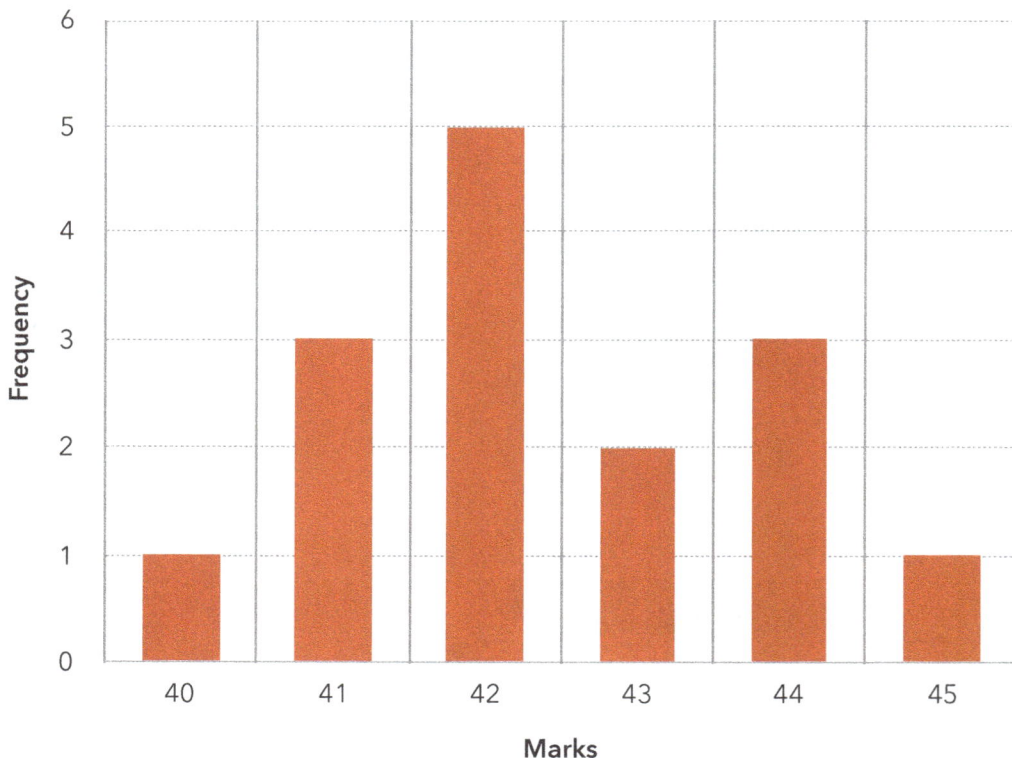

Line graphs

Line graphs are used to show trends over a period of time. For example, how the average temperature or rain fall may change from month to month over the course of a year as shown below. From the line graph it can be seen that the rainfall in the month of April measured 6 inches.

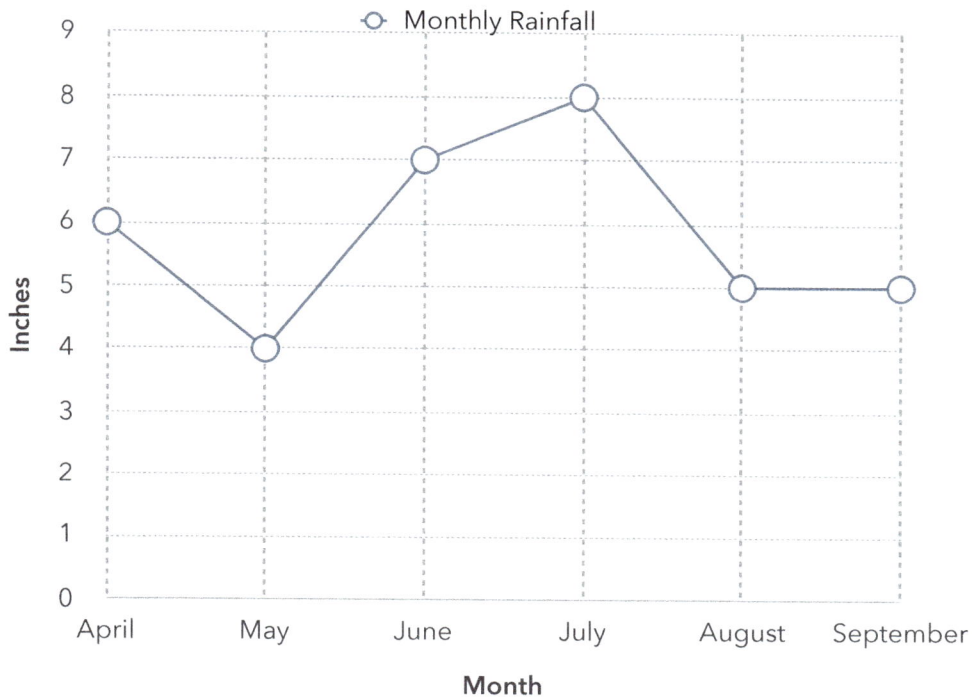

Monthly Rainfall

Example

(a) State the amount of rainfall measured in the month of May.

4 inches

(b) In which month was the greatest amount of rainfall recorded?

July

(c) Compare the rainfall recorded in the months of August and September.

The same amount of rainfall was recorded in August and September

Pie charts

A pie chart is used to show how some total amount is divided among different categories. This is done by dividing the circle into sectors. The area of each sector represents the proportion of that category which with respect to the total amount being divided. To draw a pie chart, you need to calculate the size of the angle associated with each sector involved.

$$\text{Angle at the centre of the sector} = \frac{\textit{total for that category}}{\textit{grand total}} \times 360°$$

Example

James had a party for his *7th* birthday. Twenty people went to the party of which 12 were children, 6 were women and 2 men. Draw a pie chart to show the given information.

$\textit{Angle for the children's sector} = \dfrac{12}{20} \times 360° = 216°$

$\textit{Angle for the women's sector} = \dfrac{6}{20} \times 360° = 108°$

$\textit{Angle for the men's sector} = \dfrac{2}{20} \times 360° = 36°$

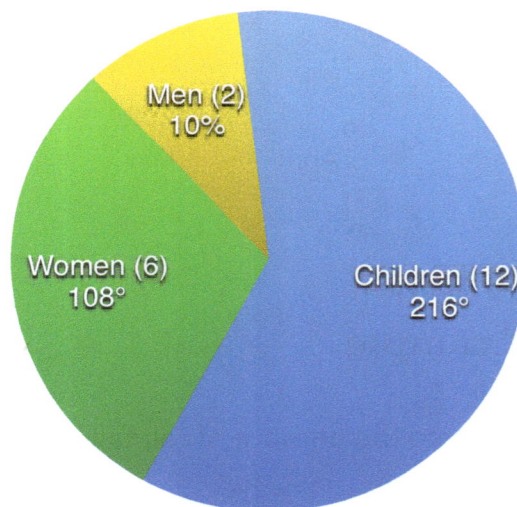

Probability

- **Probability** is the likelihood of an event occurring.
- The possible results of an experiment are called **outcomes**.

$$\textit{Probability of an event occurring} = \frac{\textit{the number of successful outcomes}}{\textit{the total number of possible outcomes}}$$

The probability of an event occurring lies between 0 and 1 is written as $\left(0 \leq P(X) \leq 1\right)$.

- If an event is impossible it will have a probability of 0.
- If the event is a certainty, the probability will be 1.

Example

1.   What is the probability of obtaining a 2 when you toss a fair die?

$$probability = \frac{number\ of\ ways\ of\ getting\ a\ 2}{total\ number\ of\ possible\ outcomes} = \frac{1}{6}$$

2.   What is the probability that the outcome will be odd?

$$probability = \frac{number\ of\ ways\ of\ getting\ an\ odd\ number}{total\ number\ of\ possible\ outcomes} = \frac{3}{6} = \frac{1}{2}$$

3.   A bag contains 4 red balls, 3 blue balls and 5 white balls. What is the probability that a ball chosen at random is

(a) blue?

$$probability\ that\ the\ ball\ is\ blue = \frac{number\ of\ blue\ balls}{total\ number\ of\ balls} = \frac{3}{12} = \frac{1}{4}$$

(b) not red?

$$probabilty\ that\ the\ ball\ is\ not\ red = \frac{number\ of\ balls\ that\ are\ not\ red}{total\ number\ of\ balls} = \frac{3+5}{12} = \frac{8}{12} = \frac{2}{3}$$

or equivalently

$$probabilty\ that\ the\ ball\ is\ not\ red = 1 - probabilty\ that\ the\ ball\ is\ red$$
$$= 1 - \frac{4}{12} = \frac{8}{12} = \frac{2}{3}$$

(c) red or white?

$$\text{probability that the ball is red or white} = \frac{\text{the number of red or white balls}}{\text{total number of balls}}$$

$$= \frac{4+5}{12} = \frac{9}{12} = \frac{3}{4}$$

> Note that the sum of the probabilities from the three possible outcomes is 1. The probability of blue is 1/4, while the probability of red or white is 3/4. Thus 1/4 + 3/4 = 1.

4.   Children in a class were asked how many siblings they had. The following table shows the results of the survey.

NUMBER OF SIBLINGS	0	1	2	3
FREQUENCY	6	17	5	2

(a) How many children were surveyed?

Total number of children = sum of frequencies = 6+17+5+2 = 30

(b) What is the probability that a student chosen at random from this class has two siblings?

$$\text{probability of having two siblings} = \frac{\text{number of children with two siblings}}{\text{total number of children}}$$

$$= \frac{5}{30} = \frac{1}{6}$$

(c) What is the probability that a student chosen at random from this class has no more than one siblings?

$$\text{probability of no more than one sibling} = \frac{\text{number of children with zero or one sibling}}{\text{total number of children}}$$

$$= \frac{6+17}{30} = \frac{23}{30}$$

Grouped data

When dealing with large amounts of data, forming an ungrouped frequency table may become tedious and the analysis of such a table may be meaningless. It is useful to group the data into **classes**. This will make analysis and comparison of the values more meaningful.

When dealing with continuous data, any whole number value actually represents a range of values. For instance, the number 25 represents any quantity within the range 24.5 to 25.5. For example, 24.7, 24.91, 25.104, and 25.43 are all approximately equal to 25, when rounded off to the nearest whole number.

When data is grouped, the whole number values that begin and end the groups are called **limits**. But since the values represent a range of values, the concept of **class boundaries** is introduced. The class boundaries include the range of values that are approximately equal to the class limits.

The following data will be used throughout this section to illustrate the various concepts. Below is a grouped frequency table showing the heights, measured in centimetres, of the basketball players at a secondary school.

Height (cm)	160-169	170-179	180-189	190-199	200-209
FREQUENCY	3	5	14	4	2

If we examine the second class (i.e. second column of data highlighted in red) in the above table, the **upper limit** of this class would be 179 and the **lower limit** is 170. However, there are a set of values that if we were to approximate, would belong to this class as well. Therefore the **upper boundary** of this class would be 179.5 and the **lower boundary** is 169.5. These can sometimes be referred to as the true limits of a class.

When analyzing a grouped frequency table, the individual values of the raw data is lost. We do not know the actual heights of each of the individuals in the classes. So we use the midpoints of each class as an approximate value for all of the items in that class. The midpoints are calculated by adding the upper and lower boundaries or upper and lower limits and dividing by 2. So the midpoint of the last class would be 204.5 because $\left(\dfrac{200+209}{2}=204.5\right)$ or $\left(\dfrac{199.5+209.5}{2}=204.5\right)$

The **class width** for each group can be found by subtracting the lower boundary from the upper boundary. The class width of the first class is 10 because $(169.5 - 159.5 = 10)$. This is true for all of the other classes in this table as well.

Example

The table below shows the times taken to run 100 metres by a sample of students. Use the grouped frequency table to answer the following questions.

Time (s)	10-12	13-15	16-18	19-21
FREQUENCY	6	12	7	2

(a) If a student ran 12.67 seconds, into which class would he be placed?

A student who runs 12.67 seconds would be placed in the 13- 15 class.

(b) What is the lower boundary of the third class?

The lower boundary of the third class is 15.5 seconds.

(c) What is the upper limit of the first class?

The upper limit of the first class is 12 seconds.

(d) Calculate the midpoint of the last class.

$$midpoint\ of\ last\ class = \frac{upper\ boundary + lower\ boundary}{2}$$

$$= \frac{21.5 + 18.5}{2} = 20\ seconds$$

(e) Calculate the class width of the second class.

$$Class\ width\ of\ second\ class = upper\ boundary - lower\ boundary$$

$$= 15.5 - 12.5 = 3\ seconds$$

Histograms

A histogram is a graphical representation of a frequency distribution for continuous data. The width of each rectangle represents the range of each class. The height of each rectangle represents the frequency for that respective class. Unlike bar charts, the bars of histograms can have varying widths and the bars **do** touch, showing the continuous nature of the data. Ideally each bar should begin and end on its respective upper and lower boundaries.

Example

Using the data on the secondary school's basketball team presented in the table below, construct a histogram to show the frequency distribution of the heights of the basketball players.

Height (cm)	160-169	170-179	180-189	190-199	200-209
FREQUENCY	3	5	14	4	2

Histogram

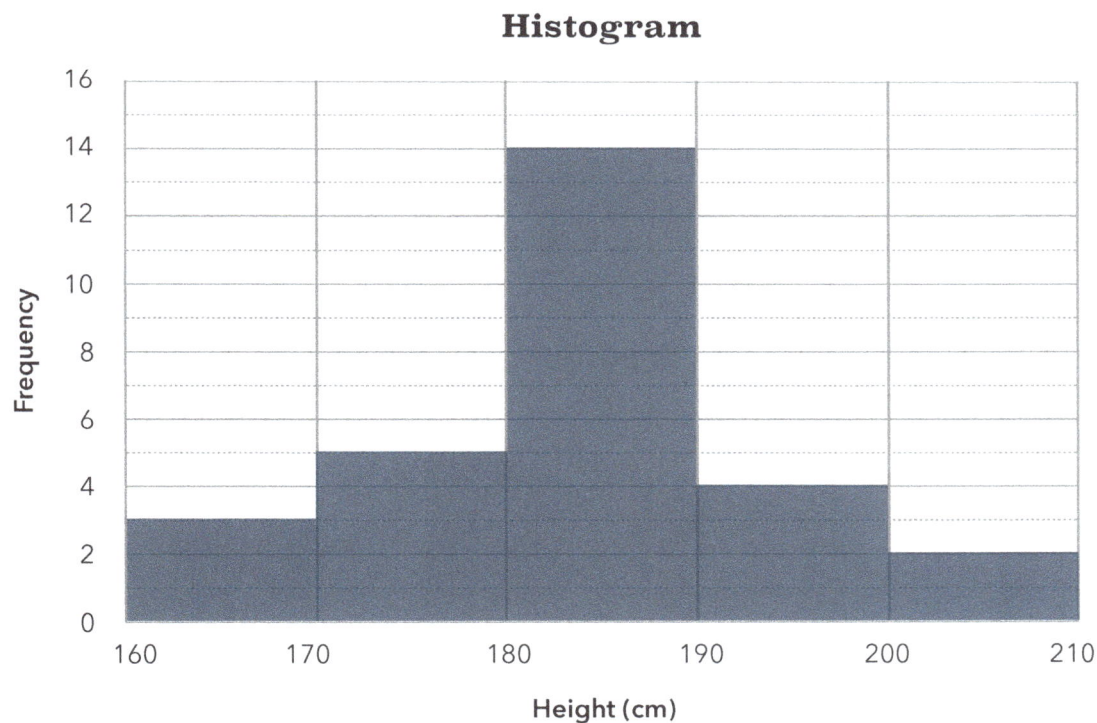

Frequency polygons

Frequency polygons can also be formed from grouped data. In this case, the frequency is plotted against the midpoint for each class. Then straight lines are drawn to join these points. A frequency polygon should start on the midpoint of the previous class and end on the midpoint of the subsequent class to ensure that the polygon is closed.

Example

Using once again the data on the secondary school's basketball team, plot the corresponding frequency polygon.

Height (cm)	Frequency (f)	Midpoints (x)
Previous	0	154.5
160-169	3	164.5
170-179	5	174.5
180-189	14	184.5
190-199	4	194.5
200-209	2	204.5
Subsequent	0	214.5
Totals	$\sum f = 28$	

Frequency Polygon

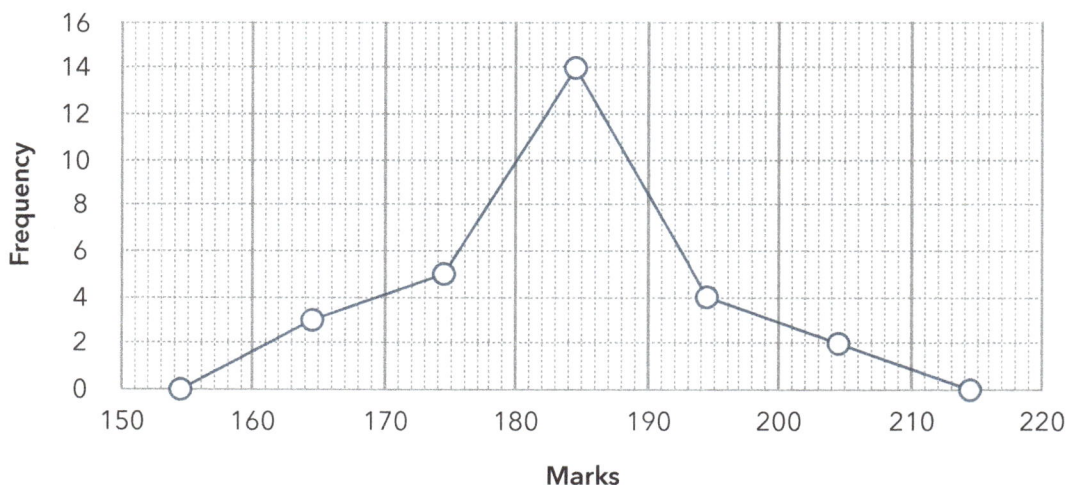

The Modal Class

The modal class of a set of data is the group with the highest frequency. It is represented on a histogram as the group with the tallest bar, or on a frequency polygon as the class with the tallest midpoint.

Example

The modal class for the set of data in the previous "Frequency Polygon" graph is 180-189 cm.

The mean of grouped data

As seen previously, the mean of a set of data is calculated by finding the total of all of the values in the data set and dividing this total by the number of values in the set. However, when data is grouped, the individual values required are not present. So assumptions need to be made. We assume that all of the items represented in a class, take on the value of the midpoint of that class, i.e. we would assume that all five persons in the 170-179 cm group are all 174.5 cm tall. Therefore, the total height of all of the persons in this group can be estimated to be 5 x 174.5 cm = 872.5 cm. This process is repeated for each of the classes. These calculations can be easily shown in a table and the mean of the data is found by using the formula, Mean = $\dfrac{\sum fx}{\sum f}$, where f is the frequency of each class and x is the corresponding midpoints of the classes

Height (cm)	Frequency (f)	Midpoints (x)	fx
160-169	3	164.5	493.5
170-179	5	174.5	872.5
180-189	14	184.5	2583
190-199	4	194.5	778
200-209	2	204.5	409
Totals	$\sum f$ =28		$\sum f$ =5136

$$\text{Mean} = \frac{\sum fx}{\sum f} = \frac{5136}{28} = 183.43 \text{ to 2 d.p.}$$

Example

A class of 32 students participated in a Mathematics test. The scores of the students is shown in the frequency table below.

Marks	Frequency (f)	Midpoint (x)	fx
40-44	4	42	168
45-49	2	47	94
50-54	6		
55-59	5		
60-64	7		
65-69	6		
70-74	2		

(a)   Copy and complete the table.

Marks	Frequency (f)	Midpoint (x)	fx
40-44	4	42	168
45-49	2	47	94
50-54	6	52	312
55-59	5	57	285
60-64	7	62	434
65-69	6	67	402
70-74	2	72	144

(b)   Calculate the mean score of the students on the test.

$$\text{Mean} = \frac{\sum fx}{\sum f}$$

$$= \frac{168 + 94 + 312 + 285 + 434 + 402 + 144}{32}$$

$$= \frac{1839}{32}$$

$$= 57.47 \text{ to 2 d.p.}$$

(c)   What is the modal class of the data shown.

The modal class is the class with the highest frequency. Therefore the modal class is 60-64.

(d)   Draw a frequency polygon to represent the data.

A frequency polygon is a plot using the midpoints of the classes against the frequency of the corresponding classes.

> Remember to include midpoints of the "previous" and "subsequent" classes in order to ensure that the polygon is closed.

Frequency Polygon

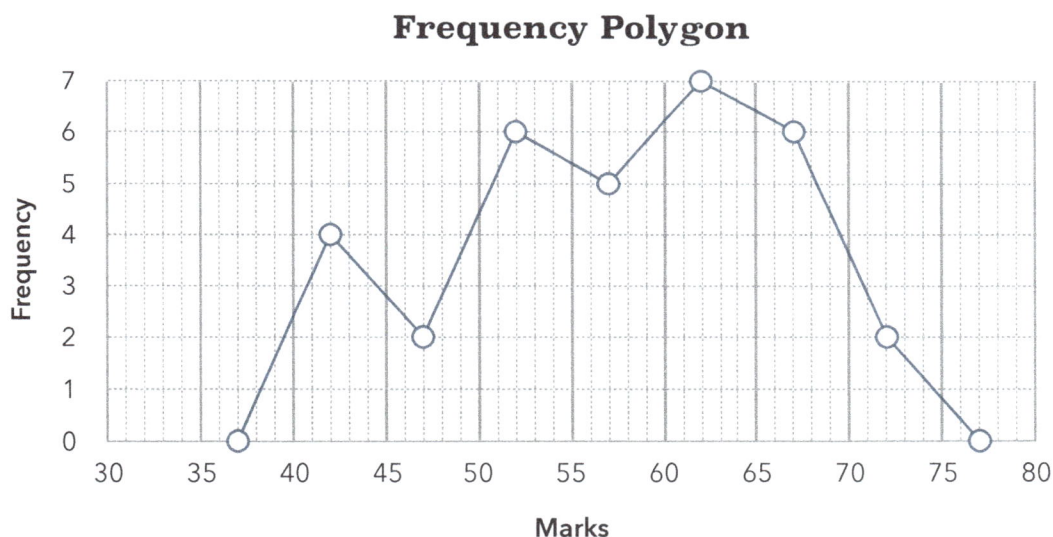

(e)   The pass mark for this test was 55 marks. What is the probability that a student chosen at random from this class has failed the test.

$$\text{probability that a student fails} = \frac{\text{number of students who failed}}{\text{total number of students}}$$

$$= \frac{4+2+6}{32}$$

$$= \frac{12}{32} = \frac{3}{8}$$

Cumulative frequency

The cumulative frequency of a class is the total of the frequencies of that class and all of the classes before it. This statistic tells us the number of items that take on the value of the upper boundary of that class or less. The following table shows the cumulative frequency for basketball players height. We shall use this example to explain various features.

Height (cm)	Frequency	Cumulative frequency
160-169	3	3
170-179	5	8
180-189	14	22
190-199	4	26
200-209	2	28

From the table, we can see that the cumulative frequency of the third class is 22. This is obtained by adding the 3 from the first class, the 5 from the second and the 14 from the third class $(3+5+14=22)$. This result tells us that there are 22 basketball players in this school that are 189.5 cm in height or shorter.

Cumulative frequency curves

A cumulative frequency curve is a graphical representation of the data reflected in a cumulative frequency table. It is drawn by plotting the cumulative frequency against the quantity under investigation. It is often used to estimate the number of items less than a certain value for the given quantity. When drawing the curve, first you must plot the coordinate coinciding with zero and the lower boundary of the first class. All other points will be plotted using coordinates pairing the upper boundary of each class and the corresponding cumulative frequencies:

$$(159.5, 0), (169.5, 3), (179.5, 8), (189.5, 22), \text{etc.}$$

Finally, a smooth curve is drawn through the points to complete the graph as illustrated below.

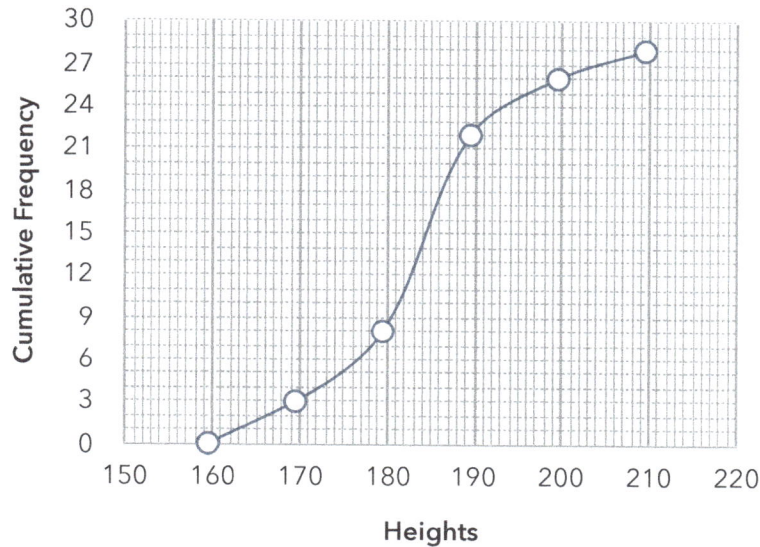

Heights

- If you want to estimate the number of players shorter than 183 cm, find 183 cm on the horizontal axis (x-axis), go up until you reach the curve, then go across to the vertical axis (y-axis) and read this value, which is 12 . Thus, 12 of the players are shorter than 183 cm.

- If you want to know how many players were taller than 192 cm, the process is the same as before. We would find 192 cm on the horizontal axis, go up until we reach the curve, then go across to the vertical axis and read this value. This time the result is approximately 24 . But we must remember that the cumulative frequency curve tells us the number of items that has this quantity or less. So, this is telling us that there are 24 players who are 192 cm or shorter. Therefore, the number of players taller than 192 cm is (28-24) = 4 players.

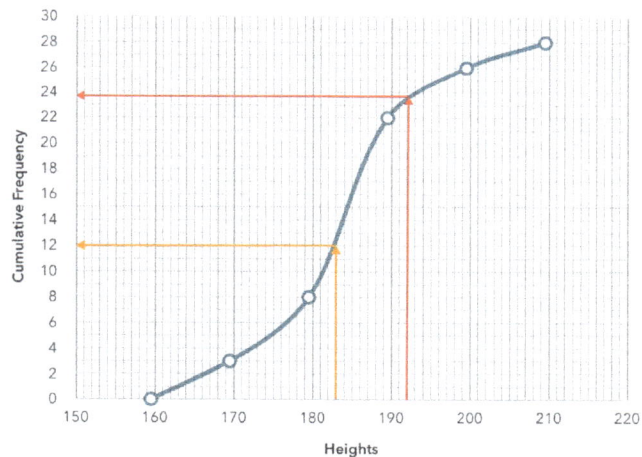

Probabilities can also be found in this manner. However, whatever value is found from the curve must be expressed as a fraction of the total number of items. For example, since there are 4 players who are taller than 192 cm, the probability that a basketball player from this school is taller than 192 cm is $\frac{4}{28}$ or $\frac{1}{7}$.

The quartiles from a cumulative frequency curve

- The **lower quartile** of a data set is the value $\frac{1}{4}$ of the way through the data once arranged in order. It is the $\left(\frac{n+1}{4}\right)^{th}$ value, often denoted as Q_1.

- You may recall that the **median** of a set of data is the value in the middle of the data set when arranged in ascending or descending order. It is the $\frac{n+1}{2}^{th}$ value, often denoted as Q_2.

- The **upper quartile** of a data set is the value $\frac{3}{4}$ of the way through the data once arranged in order. It is the $\left(\frac{3(n+1)}{4}\right)^{th}$ value, often denoted as Q_3.

In order to find these statistics you must locate the associated value on the vertical axis, go across until you reach the curve and then go to the horizontal axis, and read this value as illustrated below.

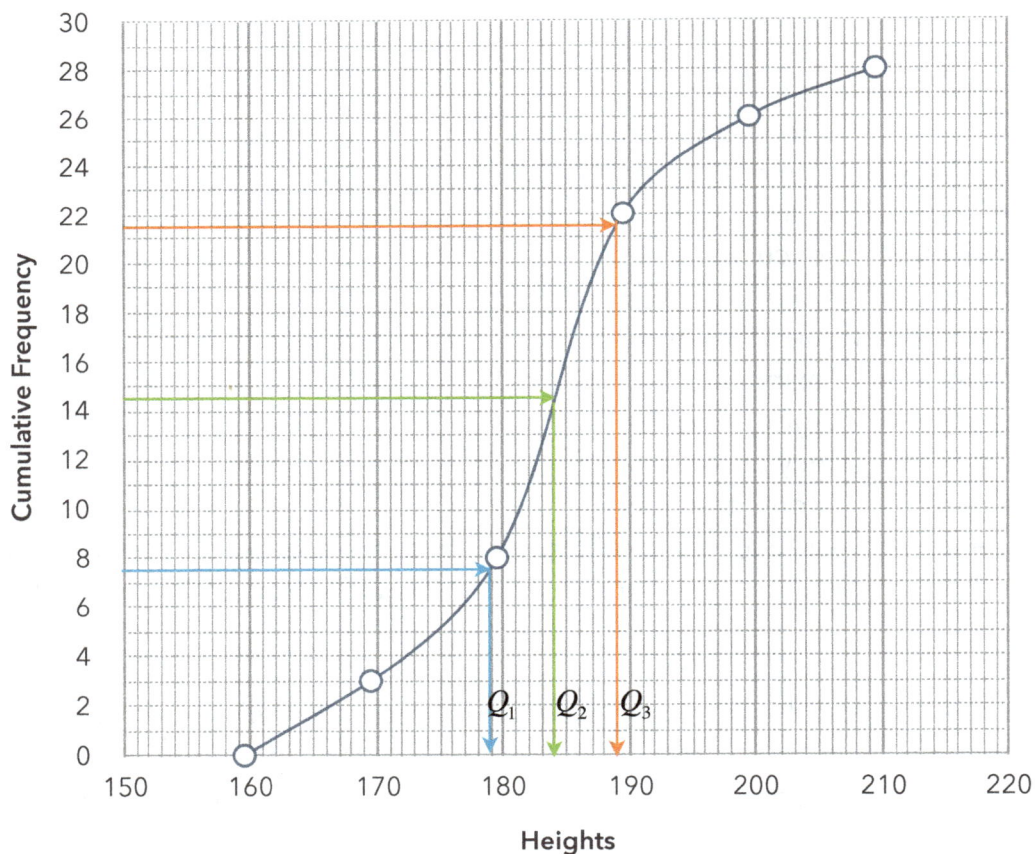

Since there are 28 basketball players, $\left(\dfrac{n+1}{4}\right)^{th}$ value would be the 7.25^{th} player.

We find 7.25 on the vertical axis, go across until we meet the curve, then go down to the horizontal axis and read this value. From the graph, we can see this is approximately 179 cm. Therefore the lower quartile Q_1 of this data set is 179 cm. This process can be repeated to obtain values for the median Q_2 and the upper quartile Q_3, which are 184 cm and 189 cm, respectively.

- The **interquartile range** is a measure of dispersion for the middle half of the data set, given by subtracting the lower quartile from the upper quartile $(Q_3 - Q_1)$

    For the data provided,
    $Interquartile\ Range = Q_3 - Q_1$
    $= 189cm - 179cm$
    $= 10cm$

- The **semi-interquartile range** is equal to half the interquartile range. So for the given example, the semi-interquartile range would be $\dfrac{10}{2} = 5$ cm.

- The **range** of a set of data is simply the highest value minus the lowest value.

Example

The table below shows the distribution of the number of fruit produced by 90 fruit trees in a particular season.

Number of fruit	Frequency	Cumulative frequency
1-10	4	4
11-20	9	13
21-30	12	
31-40	15	
41-50	22	
51-60	13	
61-70	10	
71-80	5	90

(a)   Copy and complete the table to show the cumulative frequency for the distribution.

Number of fruit	Frequency	Cumulative frequency
1-10	4	4
11-20	9	13
21-30	12	**25**
31-40	15	**40**
41-50	22	**62**
51-60	13	**75**
61-70	10	**85**
71-80	5	90

(b)   Draw a cumulative frequency curve for the number of fruit produced by the trees.

Cumulative Frequency

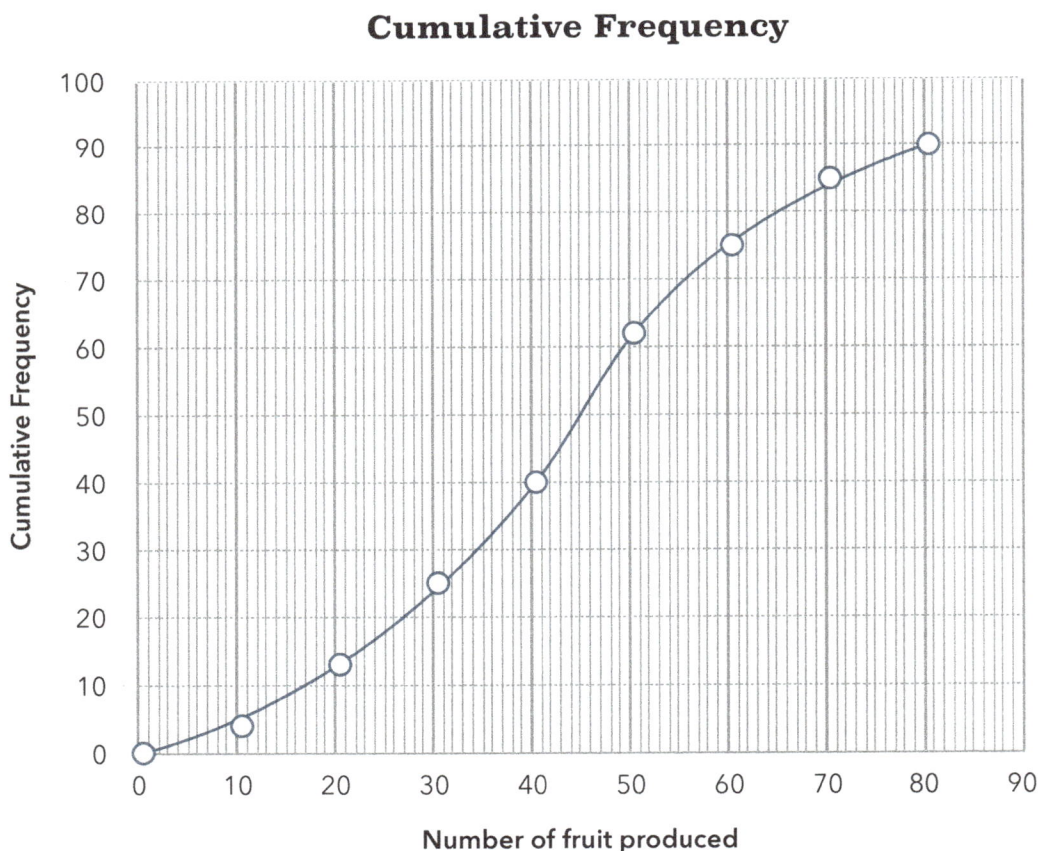

(c)   From the graph, estimate the median.

Since n is large, we can use

$$median = \frac{n}{2}^{th} \; value = \left(\frac{90}{2}\right)^{th} value = 45^{th} \; value$$

Cumulative Frequency

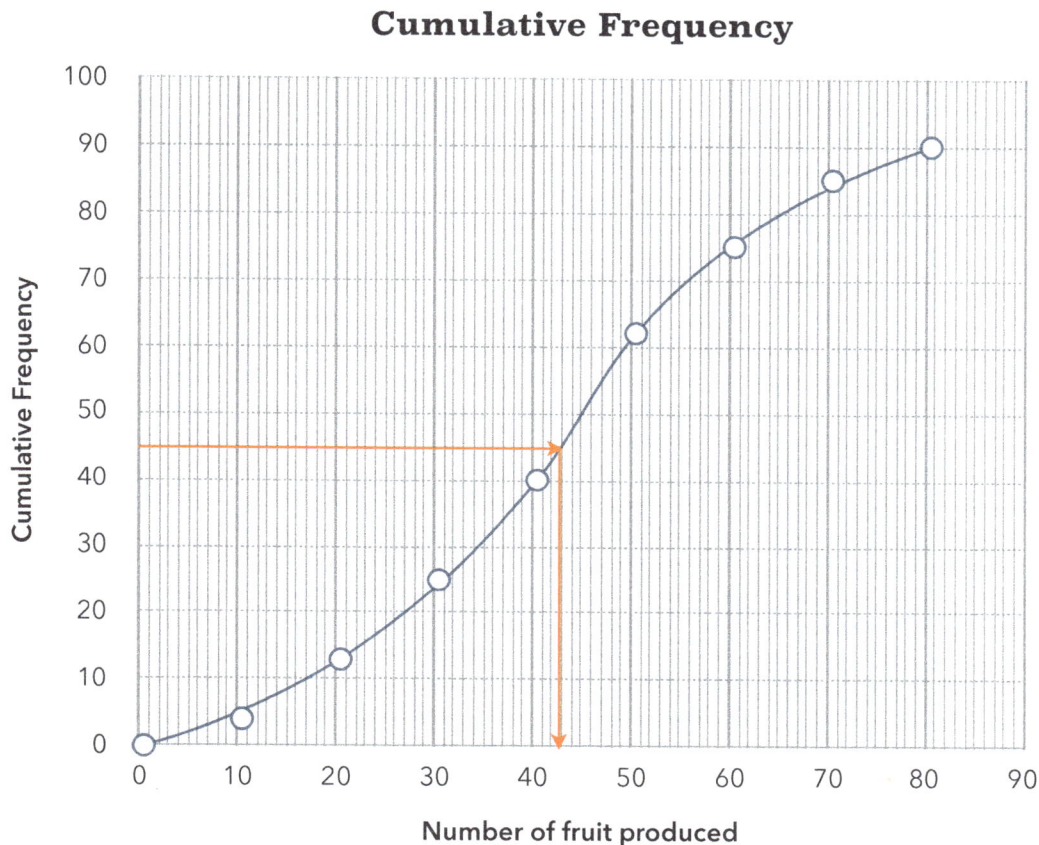

Taking the 45th value on the vertical axis, the median is approximately 43 fruit as shown in the graph above.

(d)   From the graph, estimate the inter-quartile range

$$Interquartile \; Range = Q_3 - Q_1$$

$$Lower \; quartile = \left(\frac{n}{4}\right)^{th} value$$

$$= \left(\frac{90}{4}\right)^{th} value = 22.5^{th} \; value$$

$$Upper\ quartile = \left(\frac{3n}{4}\right)^{th} value$$

$$= \left(\frac{3(90)}{4}\right)^{th} value = 67.5^{th}\ value$$

Cumulative Frequency

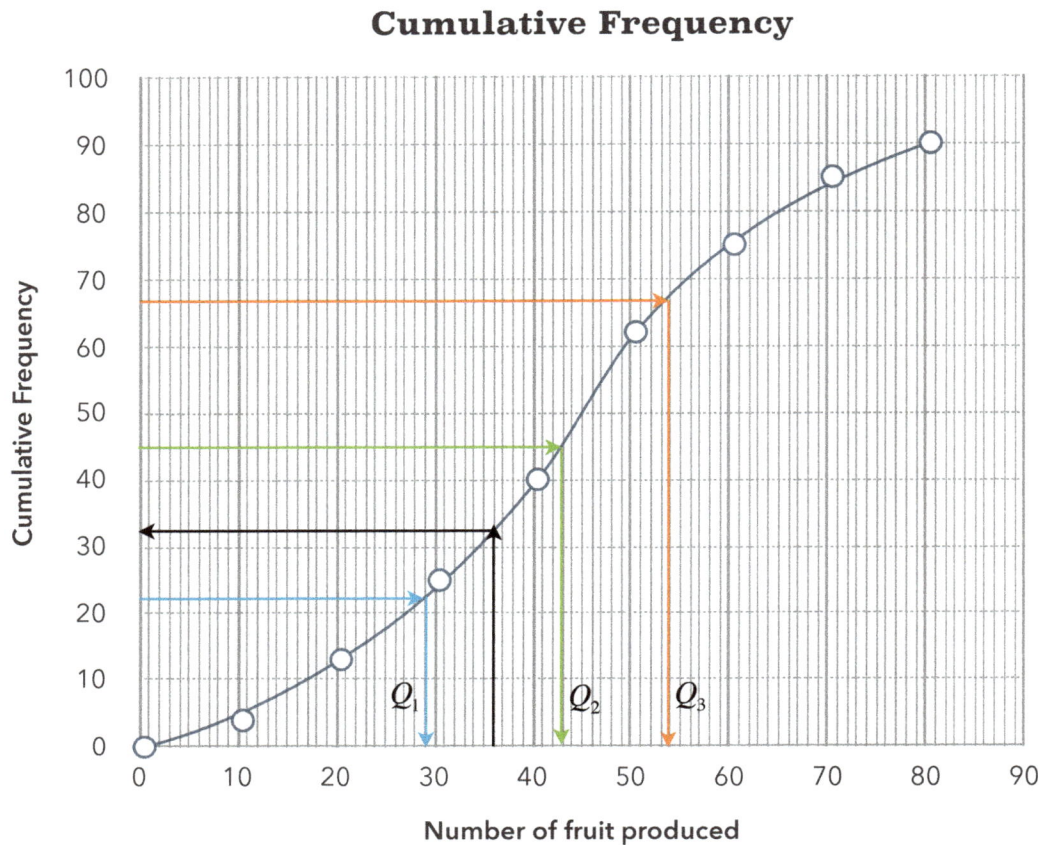

From the graph, we see that $Q_1 = 29\ fruit$ and $Q_3 = 54\ fruit$. Therefore

$Interquartile\ Range = Q_3 - Q_1$
$= 54 - 29$
$= 25\ fruit$

(e)   From the graph, estimate the number of trees that produced less than 35 fruit in the given season.

From the cumulative frequency graph above, approximately 32 trees produced less that 35 fruit in this particular season.

Questions
Statistics
Shanielle Small
Frederick Smith School
Barbados

> ▸ Video solutions to these questions via the App "**CTS Maths**" for your smartphone or tablet. Details on http://CaribbeanTeachersSeries.com

QUESTIONS

[1] The scores of 7 students in a Maths test are 4,6,7,7,3,8,7. Find the mean, median and mode of these numbers.

[2] The marks of 25 students in class are recorded.

6	6	5	7	4	5	4	3	4	6
2	7	6	5	5	5	4	6	4	7
3	5	6	4	2					

(a) Draw a frequency table.
(b) Draw a histogram representing the data.

[**3**] Sandra baked a cake using the following ingredients as represented in the pie chart below. The total mass is 432g.   What is the combined mass of flour and sugar?

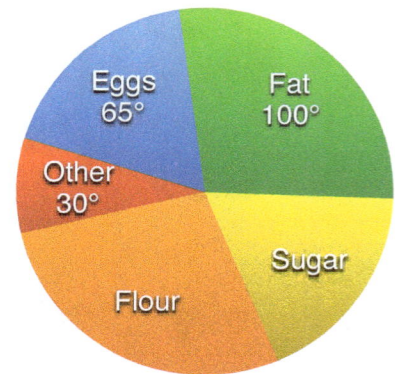

[**4**] A dice is rolled once. Calculate:
(a) Pr(5)
(b) Pr(even number)
(c) Pr(multiple of 3)
(d) Pr(number not divisible by 3)

[**5**] The below table shows the heights in cm of 15 students. Draw a frequency polygon to show this information.

Height (cm)	120-129	130-139	140-149	150-159
Number of Students	3	1	7	4

[**6**] A dice is thrown 20 times and the results are recorded in the below table.
(a) Complete the table.
(b) What is the mean score?

Score x	Frequency (f)	fx
1	2	2
2	4	
3	5	
4	3	
5	4	
6	2	

[**7**] Find the standard deviation of the eight numbers

0,5,6,6,7,8,10,14

[**8**] Here are the heights of 12 girls (cm)

134, 152, 143, 148, 159, 129, 138, 142, 137, 131, 151, 146.

Find the,
(a) lower quartile
(b) upper quartile
(c) Interquartile range

[**9**] The table below shows the scores of 50 students in a Spanish test. Draw a cumulative frequency table using the intervals 0.5 to 10.5, 10.5 to 20.5, etc.

Scores	Frequency
1-10	2
11-20	10
21-30	19
31-40	13
41-50	6

[**10**] Draw the cumulative frequency curve for the previous question.

[**11**] The marks of 50 students in a test are shown in the table below. Find the mid-interval values for each class.

Mark	Frequency (f)
0-19	4
20-39	12
40-59	21
60-79	8
80-99	5

[**12**] The percentage marks of 30 students in a Math test are shown in the table below. Write down the class boundaries for each interval.

Mark %	Number of Students
0-19	2
20-39	9
40-59	3
60-79	12
80-99	4

[**13**] A distribution of 50 scores has a mean of 25.

(a) If each score is increased by 2, find the new mean.
(b) Ten scores with a mean of 10 are removed from the original distribution. What is the mean of the remaining distribution?

[**14**] A soap manufacturer finds that in a sample of 120 soap bars, 15 are spoiled.

(a) What is the probability that a soap bar chosen at random is spoiled?
(b) How many soap bars would the manufacturer expect to be spoiled in a batch of 4000 soap bars?

[**15**] The ages of 22 boys in a sports club are shown in the table below. What is the median age?

Age	Frequency (f)
10	5
11	2
12	8
13	4
14	3

[**16**] The heights of 12 boys in centimetres (cm) are as follows:

134, 152, 143, 148, 159, 129, 138, 142, 137, 131, 151, 146

(a) Calculate the mean height of the boys.
(b) Find the range of the heights of the boys.

[**17**] The table below shows the ages of 20 children.

Age	5	6	7	8	9
Number of Children	2	4	6	3	5

(a) What is the probability that a child chosen at random is 6 years old?
(b) What is the probability that a child chosen at random is at least 8 years old?

[**18**] The bar chart below shows the shoe sizes of a sample of students.

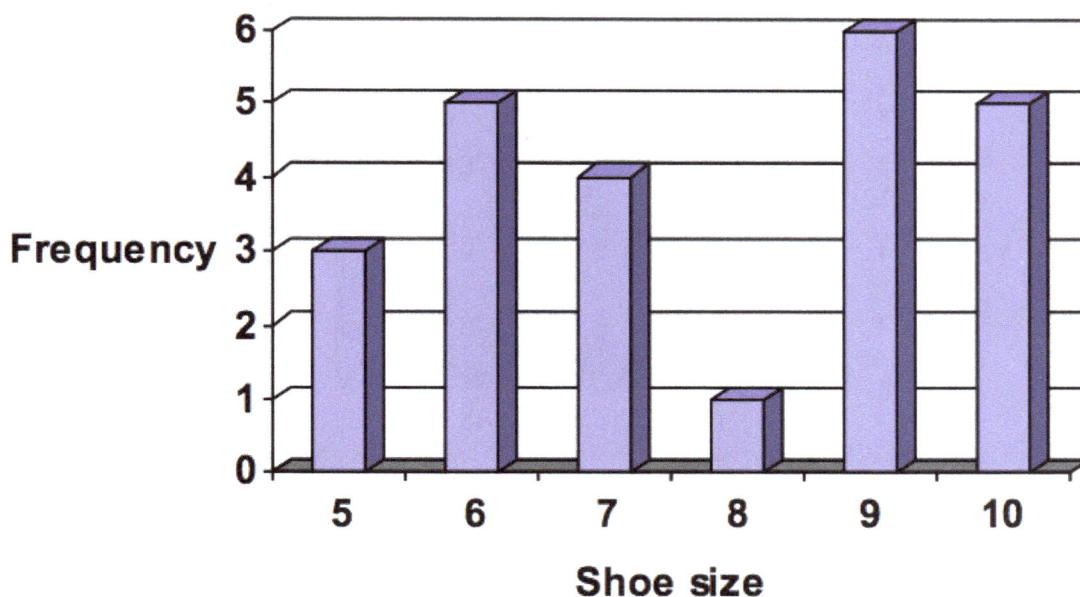

(a) What is the probability that a student chosen at random wears size 9 shoes?
(b) What is the probability that a student chosen at random has shoe size 7 or smaller?

[**19**] The pie chart below shows the percentage of land allotted to carrots, beans and lettuce. What is the size of the angle representing beans?

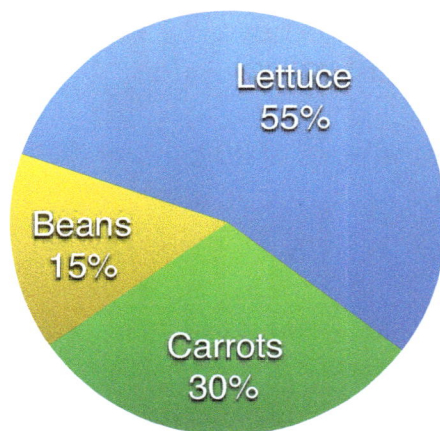

[**20**] The histogram below shows the scores of students in a Mathematics test marked out of 50.

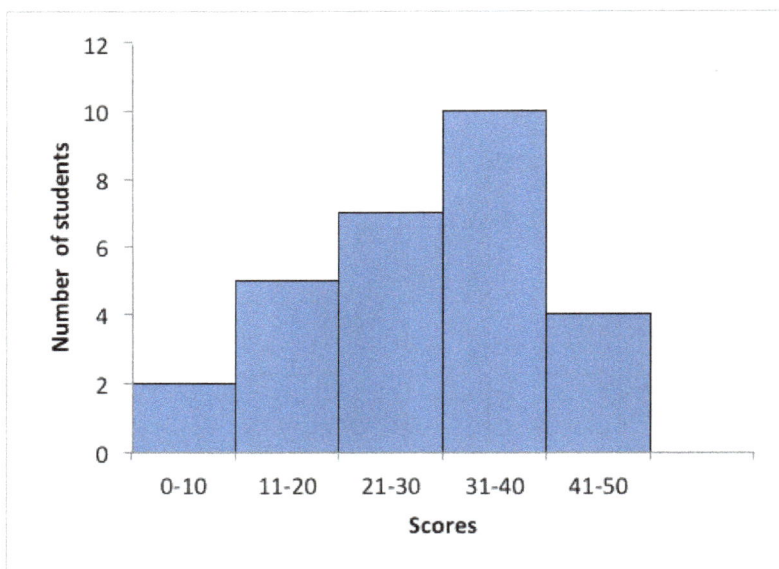

(a) How many students sat the test?

(b) What percentage of the students scored 20 marks or less?

Chapter 10
Vectors & Matrices

Richard Forde

Ellerslie School

Barbados

A matrix is a rectangular arrangement of numeric and algebraic terms and expressions.   They are used every day in fields such as computer programming, architecture, geology, statistics, surveys, economics, robotics and animation just to name a few.

In this chapter we will assume you have the prerequisite knowledge of the definitions of a matrix, the order of a matrix and the different types of matrices.

Addition and subtraction of matrices

To add/subtract two matrices, we add/subtract the numbers in corresponding positions.

$$\begin{pmatrix} a_1 & b_1 \\ c_1 & d_1 \end{pmatrix} + \begin{pmatrix} a_2 & b_2 \\ c_2 & d_2 \end{pmatrix} = \begin{pmatrix} a_1 + a_2 & b_1 + b_2 \\ c_1 + c_2 & d_1 + d_2 \end{pmatrix}$$

$$\begin{pmatrix} a_1 & b_1 \\ c_1 & d_1 \end{pmatrix} - \begin{pmatrix} a_2 & b_2 \\ c_2 & d_2 \end{pmatrix} = \begin{pmatrix} a_1 - a_2 & b_1 - b_2 \\ c_1 - c_2 & d_1 - d_2 \end{pmatrix}$$

Example

$$\begin{pmatrix} 12 & -5 \\ 3 & 6 \end{pmatrix} + \begin{pmatrix} 2 & 0 \\ 1 & -3 \end{pmatrix} = \begin{pmatrix} 12+2 & -5+0 \\ 3+1 & 6+-3 \end{pmatrix} = \begin{pmatrix} 14 & -5 \\ 4 & 3 \end{pmatrix}$$

Multiplication by a scalar

To multiply a matrix by a scalar, we multiply all terms in the matrix by the scalar quantity.

$$x \begin{pmatrix} a & b \\ c & d \end{pmatrix} = \begin{pmatrix} x \times a & x \times b \\ x \times c & x \times d \end{pmatrix} = \begin{pmatrix} ax & bx \\ cx & dx \end{pmatrix}$$

Example

$$\frac{1}{3} \begin{pmatrix} 9 & -12 \\ 0 & 4 \end{pmatrix} = \begin{pmatrix} \frac{1}{3} \times 9 & \frac{1}{3} \times -12 \\ \frac{1}{3} \times 0 & \frac{1}{3} \times 4 \end{pmatrix} = \begin{pmatrix} 3 & -4 \\ 0 & \frac{4}{3} \end{pmatrix}$$

Multiplication of two matrices

To multiply two matrices, we multiply the rows in the first matrix by the columns in the second matrix.

$$\begin{pmatrix} r_1 & r_2 \\ r_3 & r_4 \end{pmatrix} \times \begin{pmatrix} c_1 & c_3 \\ c_2 & c_4 \end{pmatrix} = \begin{pmatrix} r_1 \times c_1 + r_2 \times c_2 & r_1 \times c_3 + r_2 \times c_4 \\ r_3 \times c_1 + r_4 \times c_2 & r_3 \times c_3 + r_4 \times c_4 \end{pmatrix}$$

> Rows go from left to right and columns go from top to bottom.

> Multiplication of matrices is NOT commutative.

Example

$$\begin{pmatrix} 2 & 5 \\ 3 & 6 \end{pmatrix} \times \begin{pmatrix} 2 & 5 \\ 3 & 6 \end{pmatrix} = \begin{pmatrix} 2 \times 2 + 5 \times 1 & 2 \times 0 + 5 \times -3 \\ 3 \times 2 + 6 \times 1 & 3 \times 0 + 6 \times -3 \end{pmatrix}$$

$$= \begin{pmatrix} 4+5 & 0+(-15) \\ 6+6 & 0+(-18) \end{pmatrix} = \begin{pmatrix} 9 & -15 \\ 12 & -18 \end{pmatrix}$$

> The columns of the first matrix must be equal to the rows of the second matrix.

Determinant of a 2 x 2 matrix

The determinant of a matrix A is written as det (A), det A, or $|A|$. If the matrix does not have a name, we place the matrix between two vertical lines. For example, the determinant of $\begin{pmatrix} 2 & 5 \\ 3 & 6 \end{pmatrix}$ is written as $\begin{vmatrix} 2 & 5 \\ 3 & 6 \end{vmatrix}$.

To find the determinant of a 2 x 2 matrix we subtract the product of the minor diagonal (top right to bottom left) from the major diagonal (from top left to bottom right). Thus, if $A = \begin{pmatrix} a & b \\ c & d \end{pmatrix}$, then $\det A = ad - bc$

Examples

If the determinant of the matrix is zero then it is called a singular matrix.

1. $\begin{vmatrix} 2 & 5 \\ 3 & 6 \end{vmatrix} = (2 \times 6) - (5 \times 3) = -3$

2. The determinant of the matrix $\begin{pmatrix} 2x & x \\ 4 & 3 \end{pmatrix}$ is 18. Find the value for x.

    $\begin{vmatrix} 2x & x \\ 4 & 3 \end{vmatrix} = 6x - 4x = 2x = 18$. Therefore $x = 9$

Adjoint of a matrix

The adjoint of a matrix is found by swapping the values on the major diagonal and multiplying the values on the minor diagonal by -1.

If $A = \begin{pmatrix} a & b \\ c & d \end{pmatrix}$, then $adj(A) = \begin{pmatrix} d & -b \\ -c & a \end{pmatrix}$

Example

Find the adjoint of the matrix $B = \begin{pmatrix} 2 & -5 \\ 3 & 6 \end{pmatrix}$

$adj(B) = \begin{pmatrix} 6 & 5 \\ -3 & 2 \end{pmatrix}$

Inverse of a 2 x 2 matrix

The inverse of a 2 x 2 matrix is found by multiplying the adjoint of the matrix by the reciprocal of the determinant.

If $A = \begin{pmatrix} a & b \\ c & d \end{pmatrix}$ then $A^{-1} = \dfrac{1}{ad-bc} \begin{pmatrix} d & -b \\ -c & a \end{pmatrix}$

A matrix multiplied by its inverse gives the Identity matrix.

$$AA^{-1} = A^{-1}A = \begin{pmatrix} 1 & 0 \\ 0 & 1 \end{pmatrix}$$

Example

Given $C = \begin{pmatrix} 6 & -9 \\ 3 & 4 \end{pmatrix}$, find C^{-1}

$$C^{-1} = \frac{1}{(6\times4)-(-9\times3)} \begin{pmatrix} 4 & 9 \\ -3 & 6 \end{pmatrix}$$

A singular matrix does not have an inverse.

Specified transformation matrices

The table below shows some transformation matrices which can be used to find the vertices of an object under a specified transformation.

Matrix	Transformation
$\begin{pmatrix} 1 & 0 \\ 0 & -1 \end{pmatrix}$	Reflection in the x-axis
$\begin{pmatrix} -1 & 0 \\ 0 & 1 \end{pmatrix}$	Reflection in the y-axis
$\begin{pmatrix} 0 & 1 \\ 1 & 0 \end{pmatrix}$	Reflection in the line y=x
$\begin{pmatrix} 0 & -1 \\ -1 & 0 \end{pmatrix}$	Reflection in the line y=-x
$\begin{pmatrix} 0 & 1 \\ -1 & 0 \end{pmatrix}$	Rotation of 90 degrees clockwise (or 270 degrees anti-clockwise) about the origin
$\begin{pmatrix} -1 & 0 \\ 0 & -1 \end{pmatrix}$	Rotation of 180 degrees about the origin
$\begin{pmatrix} 0 & -1 \\ 1 & 0 \end{pmatrix}$	Rotation of 90 degrees anti-clockwise (or 270 degrees clockwise) about the origin
$\begin{pmatrix} a & 0 \\ 0 & b \end{pmatrix}$	Enlargement with the origin as centre of enlargement by a factor of a to the x-coordinate and b to the y-coordinate

Examples

1.   Use a transformation matrix to determine the coordinates of the triangle $S'T'U'$ such that $S'T'U'$ represents the image of S (-4, 3), T (-4,1) and U (-2,1) after a 90 degree clockwise rotation about the origin.

Let O represent the object STU, O' represent the image $S'T'U'$ and R represent the transformation matrix. Then

$$O = \begin{matrix} S & T & U \\ \begin{pmatrix} -4 & -4 & -2 \\ 3 & 1 & 1 \end{pmatrix} \end{matrix}$$

> For reflections in specified lines, rotations about the origin and enlargements, we multiply the translation matrix by the object matrix.
>
> **Image = Transformation x Object**

$$O' = RO$$

$$= \begin{pmatrix} 0 & 1 \\ -1 & 0 \end{pmatrix} \begin{pmatrix} -4 & -4 & -2 \\ 3 & 1 & 1 \end{pmatrix}$$

> Since matrix multiplication is not commutative the transformation matrix must come before the object matrix in the multiplication.

$$= \begin{matrix} S' & T' & U' \\ \begin{pmatrix} 3 & 1 & 1 \\ 4 & 4 & 2 \end{pmatrix} \end{matrix}$$

Therefore the coordinates are S' (3, 4), T' (1, 4) and U' (1,2)

2.   An object ABC with coordinates (1,1), (3,5) and (5,3) respectively undergoes a transformation T to produce an image $A_1B_1C_1$ with coordinates (-1, -1), (-5, -3) and (x,y).

a)   Determine the transformation matrix T used to map A to A_1, map B to B_1 and hence determine x and y.

Let X be the object and Y be the image. Then
$$Y = TX$$

Multiplying both sides by X^{-1}

$$YX^{-1} = TXX^{-1} = TI = T$$

$$Y = \begin{matrix} A & B \\ \end{matrix} \begin{pmatrix} 1 & 3 \\ 1 & 5 \end{pmatrix} \quad \text{and} \quad X = \begin{matrix} A_1 & B_1 \\ \end{matrix} \begin{pmatrix} -1 & -5 \\ -1 & -3 \end{pmatrix}$$

Therefore $X^{-1} = -\dfrac{1}{2}\begin{pmatrix} -3 & 5 \\ 1 & -1 \end{pmatrix}$

$$T = YX^{-1} = -\frac{1}{2}\begin{pmatrix} 1 & 3 \\ 1 & 5 \end{pmatrix}\begin{pmatrix} -3 & 5 \\ 1 & -1 \end{pmatrix} = -\frac{1}{2}\begin{pmatrix} 0 & 2 \\ 2 & 0 \end{pmatrix} = \begin{pmatrix} 0 & -1 \\ -1 & 0 \end{pmatrix}$$

$$C_1 = \begin{pmatrix} 0 & -1 \\ -1 & 0 \end{pmatrix}\begin{pmatrix} 5 \\ 3 \end{pmatrix} = \begin{pmatrix} -3 \\ -5 \end{pmatrix}$$

Coordinates of C_1 are (-3, -5)

b)   Describe the transformation T

Since the matrix for a reflection in the y axis is $\begin{pmatrix} 0 & -1 \\ -1 & 0 \end{pmatrix}$ then T is a reflection in

the line $y = -x$

Translation

A translation is a movement of an object through the x and/or y planes without rotation reflection or enlargement.

Example

A triangle $P(1,2), Q(1,5), R(4,2)$ undergoes a translation $M = \begin{pmatrix} 3 \\ -2 \end{pmatrix}$. Find the coordinates of the image $P'Q'R'$

$$PQR = \begin{pmatrix} 1 & 1 & 4 \\ 2 & 5 & 2 \end{pmatrix}$$

$$P'Q'R' = PQR + MMM$$

$$= \begin{pmatrix} 1 & 1 & 4 \\ 2 & 5 & 2 \end{pmatrix} + \begin{pmatrix} 3 & 3 & 3 \\ -2 & -2 & -2 \end{pmatrix} = \begin{pmatrix} 4 & 4 & 7 \\ 0 & 3 & 0 \end{pmatrix}$$

Therefore $P' = (4,2)$ $Q' = (4,3)$ and $R' = (7,0)$

Composite transformations

A composite transformation can be used to represent a sequence of transformations as a single matrix. Note that transformations are placed in the reverse order in our equation. In other words the last transformation is the first matrix in the equation.

> Matrix multiplication is associative.

Example

The object $A\,(2,-3),\,B(-2,-1),\,C(-2,-5)$ maps to the image $A'B'C'$ after a reflection in the x-axis followed by a 90 degree anticlockwise rotation about the origin. Write the transformation which maps ABC to $A'B'C'$ and hence give the coordinates of $A'B'C'$.

Let the object be P, the reflection be R_x, the rotation be R_{90} and the image be A'

$$A'B'C' = R_{90}R_x \times ABC$$

$$R_{90}R_x = \begin{pmatrix} 0 & -1 \\ 1 & 0 \end{pmatrix}\begin{pmatrix} 1 & 0 \\ 0 & -1 \end{pmatrix} = \begin{pmatrix} 0 & 1 \\ 1 & 0 \end{pmatrix}$$

$$\therefore A'B'C' = \begin{pmatrix} 0 & 1 \\ 1 & 0 \end{pmatrix}\begin{pmatrix} 2 & -2 & -2 \\ -3 & -1 & -5 \end{pmatrix} = \begin{pmatrix} -3 & -1 & -5 \\ 2 & -2 & -2 \end{pmatrix}$$

Coordinates of $A'B'C'$ are (-3, 2), (-1, -2) and (-5, -2), respectively.

Solving simultaneous equations

Matrices can also be used to solve simultaneous equations.

Example

Solve the pair of simultaneous equations $2x + y = 3$ and $x - 4y = 6$ using matrices.

Expressing the problem in matrix form gives

$$\begin{pmatrix} 2 & 1 \\ 1 & -4 \end{pmatrix}\begin{pmatrix} x \\ y \end{pmatrix} = \begin{pmatrix} 3 \\ 6 \end{pmatrix}$$

> The coefficients of x go in the first column and the coefficients of y in the second column.

Let $A = \begin{pmatrix} 2 & 1 \\ 1 & -4 \end{pmatrix}$, $Y = \begin{pmatrix} x \\ y \end{pmatrix}$ and $X = \begin{pmatrix} 3 \\ 6 \end{pmatrix}$

Thus $AY = X$

$A^{-1}AY = A^{-1}X$

$\therefore IY = A^{-1}X$

> Its important to remember that
> $$X^{-1}X = I$$

Now recall that any matrix multiplied by the identity matrix gives back that matrix.

Therefore $Y = A^{-1}X$ and so we find the inverse of $\begin{pmatrix} 2 & 1 \\ 1 & -4 \end{pmatrix}$ and multiply it by $\begin{pmatrix} 3 \\ 6 \end{pmatrix}$.

Hence $Y = A^{-1}X$

$$\begin{pmatrix} x \\ y \end{pmatrix} = -\frac{1}{9}\begin{pmatrix} -4 & -1 \\ -1 & 2 \end{pmatrix}\begin{pmatrix} 3 \\ 6 \end{pmatrix} = -\frac{1}{9}\begin{pmatrix} -18 \\ 9 \end{pmatrix} = \begin{pmatrix} 2 \\ -1 \end{pmatrix}$$

Ans: $x = 2$ and $y = -1$.

Questions
Vectors & Matrices
Richard Forde
Ellerslie School
Barbados

> ▸ Video solutions to these questions via the App "**CTS Maths**" for your smartphone or tablet. Details on http://CaribbeanTeachersSeries.com

QUESTIONS

[1] Evaluate the determinant of $\begin{pmatrix} 6 & 2 \\ 3 & 5 \end{pmatrix}$ and $\begin{pmatrix} 5 & 10 \\ 2 & 4 \end{pmatrix}$

[2] Given the singular matrix $A = \begin{pmatrix} 4 & 3 \\ b & 6 \end{pmatrix}$, find b

[3] Given that $C = \begin{pmatrix} 5 & 8 \\ -7 & -2 \end{pmatrix}$, find adjoint C

[4] If $A = \begin{pmatrix} 3 & -2 \\ 5 & 4 \end{pmatrix}$, find the inverse of A

[5] If $P = \begin{pmatrix} 3 & 2 \\ 6 & 5 \end{pmatrix}$, show that $PP^{-1} = I$

[6] Show that the matrix $\begin{pmatrix} 0 & 1 \\ -1 & 0 \end{pmatrix}$ represents a 90 degree clockwise rotation about the origin.

[7] Given that $P = \begin{pmatrix} 0 & 1 \\ -1 & 0 \end{pmatrix}$ is a transformation matrix for a 90 degree clockwise rotation about the origin, show that $Q = \begin{pmatrix} -1 & 0 \\ 0 & -1 \end{pmatrix}$ is a transformation matrix representing a rotation of 180 degrees about the origin.

[8] Use a transformation matrix to determine the coordinates of $S'T'U'$ such that $S'T'U'$ represent the images of STU after a 90 degree rotation about the origin.

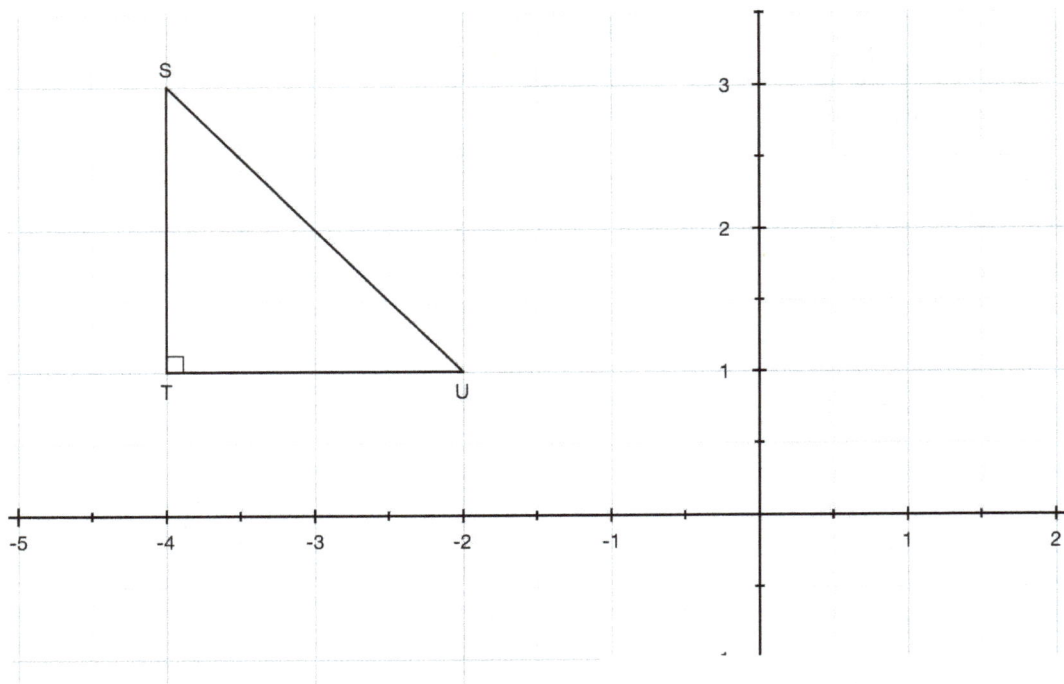

[9] Determine a matrix G that represents a 90 degree anti-clockwise rotation of the points D and B about the origin.

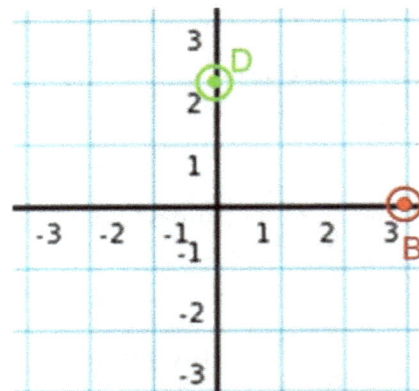

[**10**] An object *ACEG* has coordinates *A* (1, 3), *C* (3, 6), *E* (2, 6) and *G* (3,1). If *ACEG* maps to *A'C'E'G'* under a 90 degree anti-clockwise rotation about the origin, give the coordinates of *A'*, *C'*, *E'* and *G'*.

[**11**] The object *DAWN* undergoes a transformation which maps into *D' A' W' N'*. Describe clearly the type of transformation and write the matrix which represents the transformation.

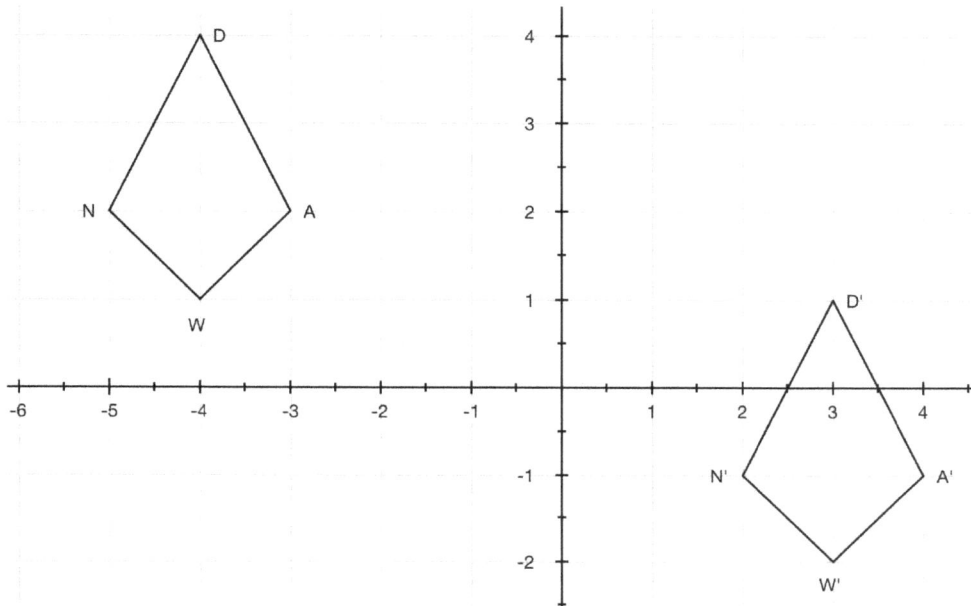

[**12**] The diagram below shows a reflection in the y axis, mapping the object *STU* to its image *S' T' U'* . Use the diagram below to determine the transformation matrix for the reflection of an object in the y axis.

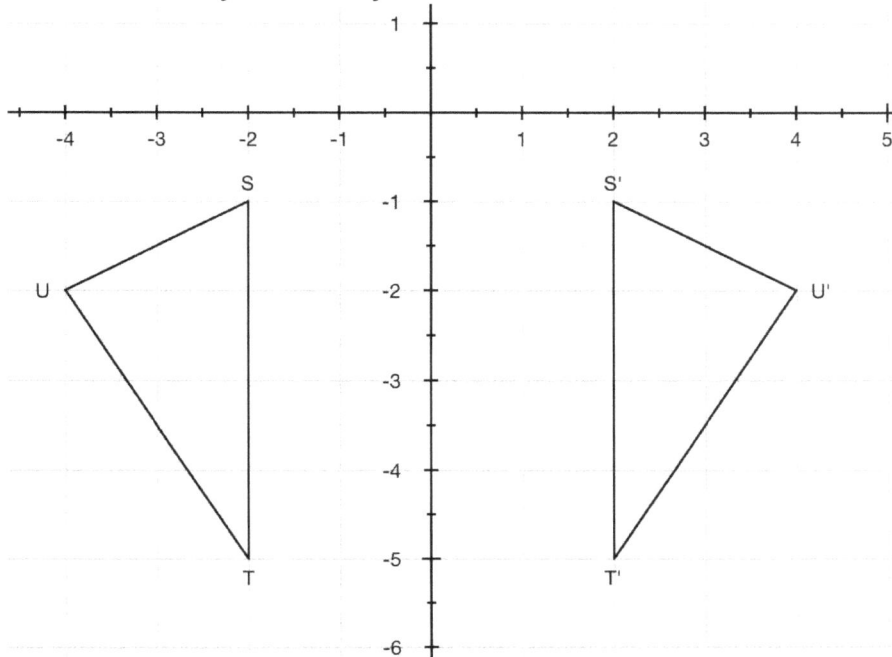

[**13**] The coordinates of AGX are A (2,3), G (8,3) and X (2,10). AGX is reflected on the x-axis giving the image A' G' X' with Coordinates A' (2, -3), G' (8, -3) and X' (x, y). Use the information above the determine the matrix R use to map A to A' and G to G' and hence, using R, find X and Y.

[**14**] The diagram below shows two points A (0, 4), D (-4, 0) and the line y=x. Using the two points A and D and their images A' and D' respectively, show that the matrix R, associated with this transformation is $\begin{pmatrix} 0 & 1 \\ 1 & 0 \end{pmatrix}$.

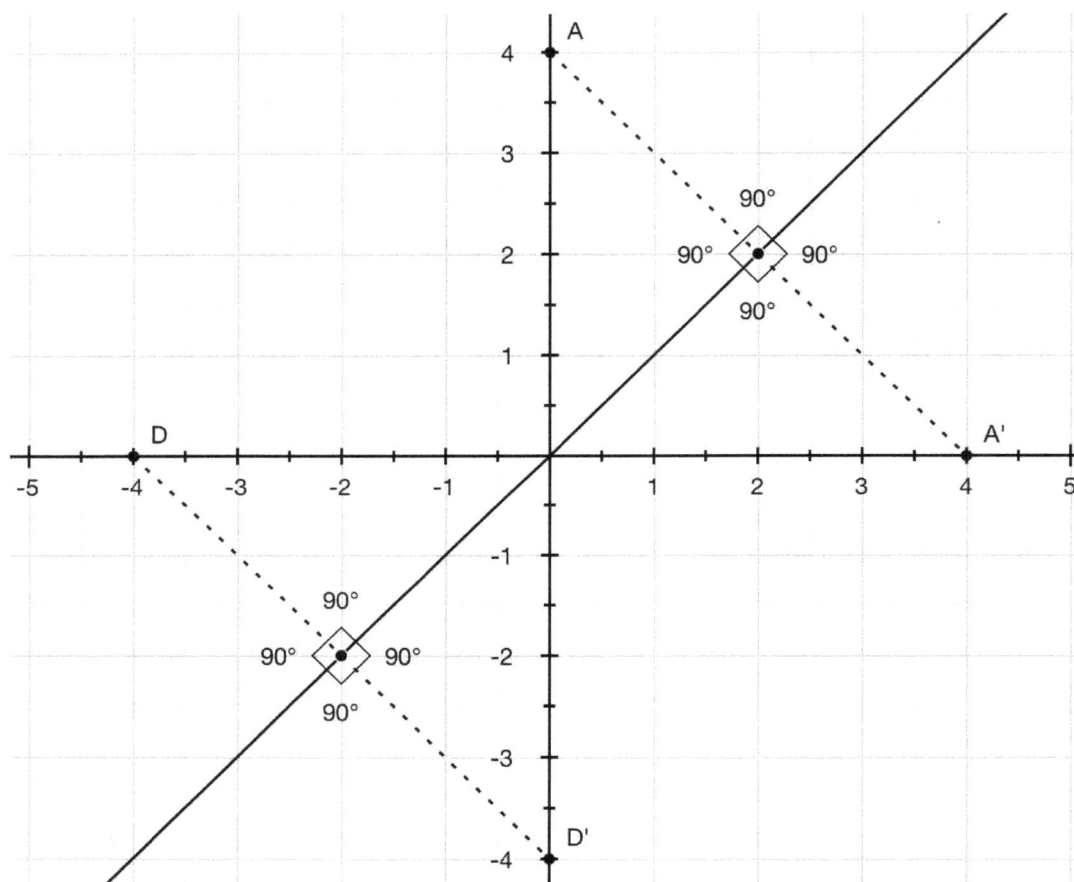

[**15**] An object ABCD is shown in the graph. Given that the matrix $J = \begin{pmatrix} 0 & 1 \\ 1 & 0 \end{pmatrix}$, is a reflection in the line $y = x$, find:

a)   the matrix K which represents a reflection in the line $y = -x$.
b)   the coordinates of the image of ABCD for a reflection in the line $y = -x$.

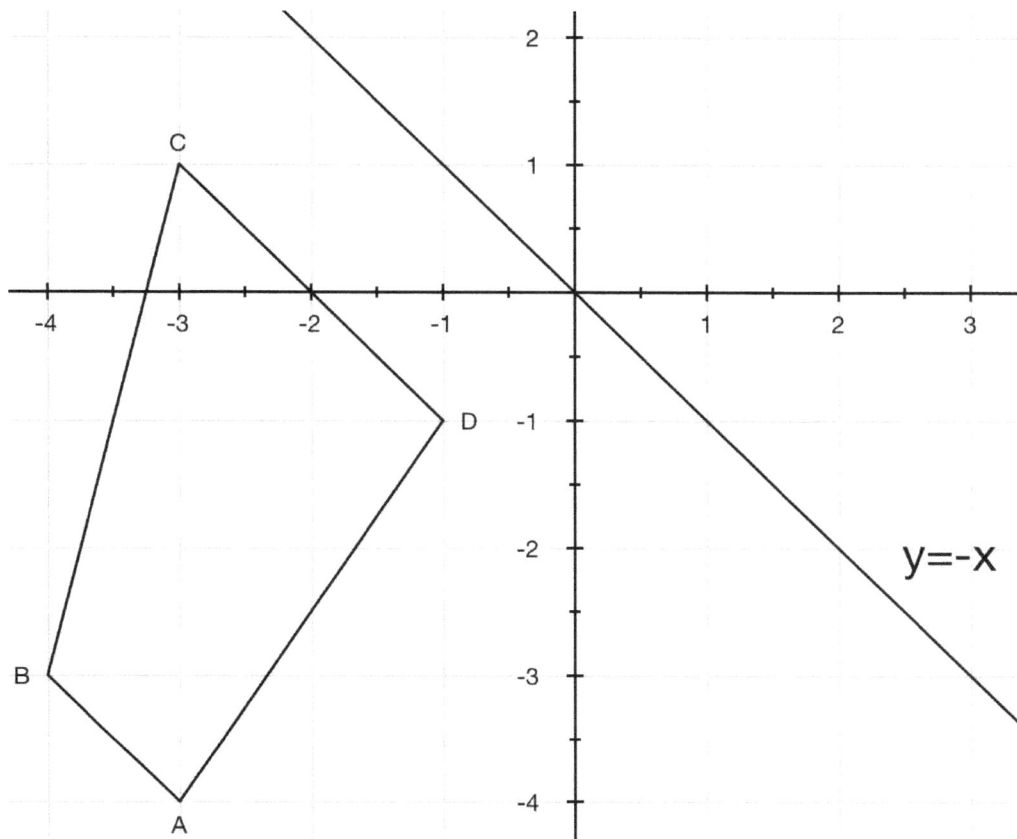

[**16**] The triangle XYZ with vertices X (3,4), Y (6,6) and Z (3,6) maps to its image X′ Y′ Z′ under an enlargement with centre at the origin and scale factor p. If X′ is (9,12) find the:

a)   scale factor p
b)   transformation matrix E for the enlargement
c)   coordinates of Y′ and Z′

[**17**] An object MAN reflects in the line $y = x$ to form the image M′ A′ N′. The image M′ A′ N′ is then reflected in the x axis to give M″ A″ N″. Determine a matrix which maps MAN to M″ A″ N″. Describe fully the single transformation used to map the object MAN to its image M″ A″ N″.

[**18**] Use the diagram shown to answer the following.

a)   State what type of transformation maps STU to S′ T′ U′ and give the matrix A for this transformation,

b)   State what type of transformation maps S′ T′ U′ to S″ T″ U″, and give the matrix for this transformation

c)   Determine a single transformation Q which maps STU to S″ T″ U″.

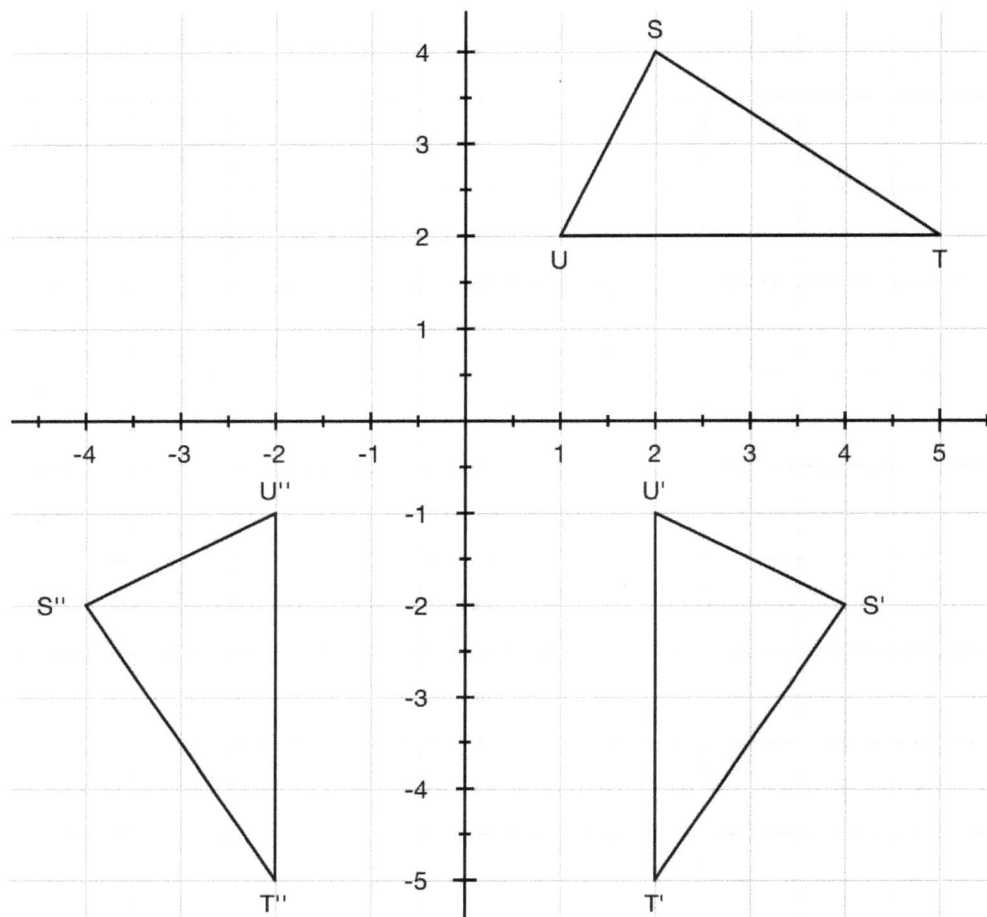

[**19**] Solve the following simultaneous equations below using matrices.

$x + 4y = -11$
$2x - y = 5$

[**20**] Find x and y in the matrix equation below.

$$x \begin{pmatrix} 2 \\ 4 \end{pmatrix} + y \begin{pmatrix} 3 \\ -3 \end{pmatrix} = \begin{pmatrix} 19 \\ -7 \end{pmatrix}$$

Feedback

Typos & Suggestions

Janak Sodha

CTS Founder &
Editor-In-Chief

UWI, Barbados

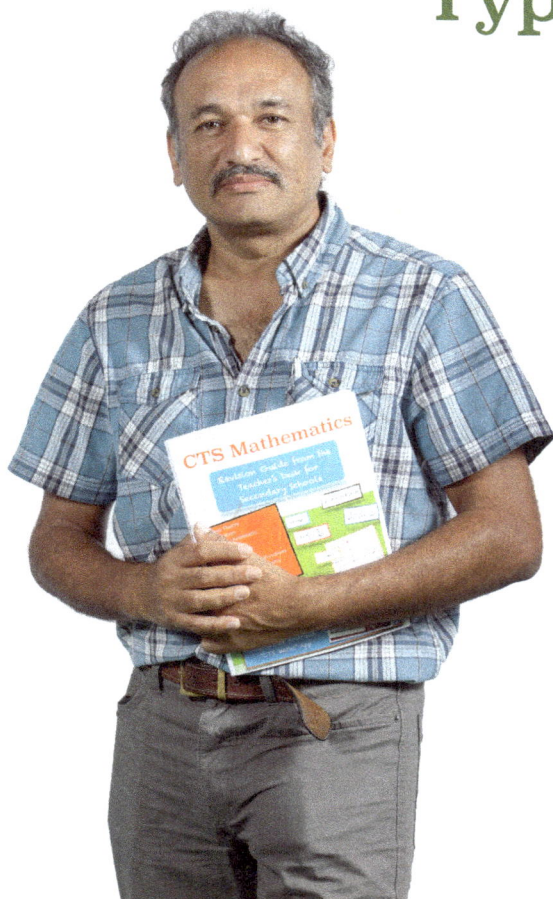

Its our first book, for our students throughout the Caribbean. We hope you find this book an invaluable resource. Please do take the time to send us feedback via the online form on the website:

http://CaribbeanTeachersSeries.com

- **Typos** - Will be listed on this website. Don't forget to indicate the page number in your reference to the typo found.

- **Suggestions** - Students, Teachers and Parents, how may we improve this book? What types of examples are missing? Your feedback will help us continue to evolve this textbook. Thank you in anticipation.

Acknowledgements

Berger Paints http://www.bergeronline.com/caribbean/	Sponsorship of iPads.
FirstCaribbean International Bank https://www.cibcfcib.com/fcib/	Sponsorship of iPads.
Sir Hilary Beckles Principal, University of the West Indies http://www.cavehill.uwi.edu	Sponsorship of iPads.
David Garner XQUIZIT Digital Photos http://www.xquizitphotos.com	Photographs of the teachers.
Dr. Anthony Fisher Director of External Relations Office of Student Corporate & Alumni Relations (OSCAR) University of the West Indies http://www.cavehill.uwi.edu	Helped to secure sponsorship for iPads.
Rashid Holder Caved Geeks http://www.cavedgeeks.com	Wrote the code for the iOS and Android CTS Maths App.
	Funding was also provided under the Barbados Human Resource Development Strategy, which is financed by the European Union

www.ingramcontent.com/pod-product-compliance
Lightning Source LLC
Chambersburg PA
CBHW061357210326
41598CB00035B/6015